IB Skills

T0173131

Sciences

A PRACTICAL GUIDE STUDENT BOOK

Fiona Clark
David Mindorff
Rita Pak
with Barclay Lelievre

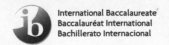

International Baccalaureate
Baccalauréat International
Bachillerato Internacional

Sciences: A practical guide (Student book)

Published on behalf of the International Baccalaureate Organization, a not-for-profit educational foundation of 15 Route des Morillons, 1218 Le Grand-Saconnex, Geneva, Switzerland by the International Baccalaureate Organization (UK) Ltd, Peterson House, Malthouse Avenue, Cardiff Gate, Cardiff, Wales CF23 8GL United Kingdom, represented by IB Publishing Ltd, Churchillplein 6, The Hague, 2517JW The Netherlands

Website: www.ibo.org

The International Baccalaureate Organization (known as the IB) offers four high-quality and challenging educational programmes for a worldwide community of schools, aiming to create a better, more peaceful world. This publication is one of a range of materials produced to support these programmes.

© International Baccalaureate Organization 2015

Published 2015

The rights of Barclay Lelievre, Fiona Clark, David Mindorff and Rita Pak to be identified as authors of this work have been asserted by them in accordance with sections 77 and 78 of the Copyright, Designs and Patent Act 1988.

All rights reserved. No part of this publication may be reproduced, stored in a retrieval system, or transmitted, in any form or by any means, without the prior written permission of the IB, or as expressly permitted by law or by the IB's own rules and policy. See http://www.ibo.org/copyright.

International Baccalaureate, **Baccalauréat International** and **Bachillerato Internacional** are registered trademarks of the International Baccalaureate Organization.

IB merchandise and publications can be purchased through the IB store at http://store.ibo.org. General ordering queries should be directed to the Sales and Marketing Department at sales@ibo.org.

British Library Cataloguing in Publication Data

A catalogue record for this book is available from the British Library

ISBN: 978-1-910160-04-6
MYP370
Typeset by Q2A Media Services Pvt Ltd
Printed and bound in Dubai.

Acknowledgments

We are grateful for permission to reprint copyright material and other content:

p7 blocks: ©iStockphoto, tools: ©iStockphoto, plans: ©iStockphoto, house: ©iStockphoto; p10 pot with lid: http://commons.wikimedia.org/wiki/File:Pot_on_stove.jpg, pot: http://commons.wikimedia.org/wiki/File:Water_boiling_in_a_pot_on_a_stove.jpg, insulated bottle: ©iStockphoto, Earth: http://commons.wikimedia.org/wiki/File:The_Blue_Marble.jpg, aquarium: http://commons.wikimedia.org/wiki/File:Aquarium-Monaco1.jpeg, Biosphere II: http://commons.wikimedia.org/wiki/File:Biosphere_2_Habitat_%26_Lung_2009-05-10.jpg; p12 Venus tablet: http://commons.wikimedia.org/wiki/File:Venus_Tablet_of_Ammisaduqa.jpg; p24 plague doctor: http://commons.wikimedia.org/wiki/File:Plague_Doctor_(The_Mirror_of_literature,_1841).png; p25 Sir Cloudesley Shovell: http://commons.wikimedia.org/wiki/File:Sir_Cloudesley_Shovell,_1650-1707.jpg; p29 Refractometer: http://commons.wikimedia.org/wiki/File:Portable-Refractometer-09.jpg; p31 Oxtant: http://commons.wikimedia.org/wiki/File:Hadley%27s_reflecting_quadrant.png; p34 Sun: https://archive.org/details/SPD-SOHO-eit001; p43 insulated mug: ©iStockphoto; p53 Darwin's finches: http://commons.wikimedia.org/wiki/File:Darwin%27s_finches_by_Gould.jpg; p73 Nikola Tesla: http://commons.wikimedia.org/wiki/File:Tesla2.jpg; p74 transformer: http://commons.wikimedia.org/

IB Publishing would like to thank Aileen Harvey and Dan Rosen for their contributions to the writing and editing of this book.

wiki/File:Power_Pole_With_Transformer,_Ponsonby.jpg, duckweed: ©Mary Wickison; p84 Linus Pauling: http://commons.wikimedia.org/wiki/Linus_Pauling#mediaviewer/File:Pauling.jpg; p86 Weddell seals: http://commons.wikimedia.org/wiki/File:Diving_weddell_seals.jpg; p90 egg shell: ©POWER AND SYRED/SCIENCE PHOTO LIBRARY; p96 scanner: ©iStockphoto; p107 statins: ©iStockphoto; p109 Leaf miner: http://commons.wikimedia.org/wiki/File:Lonicera_leaf_miner_kz.jpg; p111 disclosing tablet: ©iStockphoto; p127 genetically modified maize: ©iStockphoto; p129 map of PM10 particulates: World Health Organization, Global Health Observatory, http://gamapserver.who.int/mapLibrary/Files/Maps/Global_pm10_cities.png, ©WHO; p133 Hydrogen Ion Concentration map: ©NADP [National Atmospheric Deposition Program], acid rain effects: ©iStockphoto; p143 microscope: http://commons.wikimedia.org/wiki/File:Hooke_Microscope.jpg; p146 feathers: http://upload.wikimedia.org/wikipedia/commons/2/28/Types_of_feathers.jpg; p148 aloe plant: http://commons.wikimedia.org/wiki/File:Aloe_vera_plant.JPG; p157 recycling: ©iStockphoto, potter's wheel: ©iStockphoto; p167 woodlice: http://upload.wikimedia.org/wikipedia/commons/b/be/Woodlice_02.JPG; p168 swimming paths: Source: D W Sims and V A Quayla (1998) Nature 393, pages 460–464, mimosa pudica: http://commons.wikimedia.org/wiki/File:Mimosa_pudica_open.JPG; p169 winged acer seed: http://commons.wikimedia.org/wiki/File:Acer_saccharinum_seeds.jpg; p182 gecko foot: ©POWER AND SYRED/SCIENCE PHOTO LIBRARY, gecko foot hairs: ©POWER AND SYRED/SCIENCE PHOTO LIBRARY; p184 diagnostic tool: ©iStockphoto; p190 antacid tablets and liquids: ©iStockphoto; p202 Robert Millikan: http://commons.wikimedia.org/wiki/File:Robert_Andrews_Millikan.jpg; p204 flowers to fruit: ©CLAUS LUNAU/SCIENCE PHOTO LIBRARY; p222 key bacterial genus: http://www.nature.com/nature/journal/v473/n7346/full/nature09944.html, Nature, Vol. 472 No. 7343, April 21, 2011; p224 lichen: ©iStockphoto; p232 Chernobyl's exclusion zone: http://commons.wikimedia.org/wiki/File:Red_Forest_Hill.jpg; p234 dive depths: modified from 41 Heart Rates of Northern Elephant Seals Diving at Sea and Resting on the Beach. R. D. Andrews and others, 1997. Journal of Experimental Biology, 200, 2083-2095. Used with permission.

The IB may use a variety of sources in its work and checks information to verify accuracy and authenticity, particularly when using community-based knowledge sources such as Wikipedia. The IB respects the principles of intellectual property and makes strenuous efforts to identify and obtain permission before publication from rights holders of all copyright material used. The IB is grateful for permissions received for material used in this publication and will be pleased to correct any errors or omissions at the earliest opportunity.

This publication may contain web addresses of websites created and maintained by other public and/or private organizations. The IB provides links to these sites for information purposes only. The presence of a link is not an IB endorsement of the site and the IB is not responsible for the content of external internet sites. When you copy a link to an outside website(s), you are subject to the privacy and security policies of the owners/sponsors of the outside website(s). The IB is not responsible for the information collection practices of non-IB sites. You are reminded that while all web addresses referenced in this publication were correct at the time of writing, links can quickly become obsolete and we are unable to guarantee that links will continue to remain active. We apologize if you find that a particular link no longer works.

IB learner profile

The aim of all IB programmes is to develop internationally minded people who, recognizing their common humanity and shared guardianship of the planet, help to create a better and more peaceful world.

As IB learners we strive to be:

INQUIRERS
We nurture our curiosity, developing skills for inquiry and research. We know how to learn independently and with others. We learn with enthusiasm and sustain our love of learning throughout life.

KNOWLEDGEABLE
We develop and use conceptual understanding, exploring knowledge across a range of disciplines. We engage with issues and ideas that have local and global significance.

THINKERS
We use critical and creative thinking skills to analyse and take responsible action on complex problems. We exercise initiative in making reasoned, ethical decisions.

COMMUNICATORS
We express ourselves confidently and creatively in more than one language and in many ways. We collaborate effectively, listening carefully to the perspectives of other individuals and groups.

PRINCIPLED
We act with integrity and honesty, with a strong sense of fairness and justice, and with respect for the dignity and rights of people everywhere. We take responsibility for our actions and their consequences.

OPEN-MINDED
We critically appreciate our own cultures and personal histories, as well as the values and traditions of others. We seek and evaluate a range of points of view, and we are willing to grow from the experience.

CARING
We show empathy, compassion and respect. We have a commitment to service, and we act to make a positive difference in the lives of others and in the world around us.

RISK-TAKERS
We approach uncertainty with forethought and determination; we work independently and cooperatively to explore new ideas and innovative strategies. We are resourceful and resilient in the face of challenges and change.

BALANCED
We understand the importance of balancing different aspects of our lives—intellectual, physical, and emotional—to achieve well-being for ourselves and others. We recognize our interdependence with other people and with the world in which we live.

REFLECTIVE
We thoughtfully consider the world and our own ideas and experience. We work to understand our strengths and weaknesses in order to support our learning and personal development.

The IB learner profile represents 10 attributes valued by IB World Schools. We believe these attributes, and others like them, can help individuals and groups become responsible members of local, national and global communities.

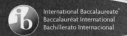
International Baccalaureate®
Baccalauréat International
Bachillerato Internacional

© International Baccalaureate Organization 2013
International Baccalaureate® | Baccalauréat International® | Bachillerato Internacional®

Contents

How to use this book

As well as introducing you to the 3 key concepts and 12 of the related concepts in the Middle Years Programme (MYP) sciences course, this book will also help you practise all the skills you need to reach the highest level of the MYP assessment criteria.

This book has been divided into chapters on key and related concepts. Throughout the book you will find features that will help you link your learning to the core elements of the MYP.

On the first page of each of the related concept chapters you will find:

- the topics you will be focusing on
- the inquiry questions you will be considering
- a checklist of skills you will practice
- a glossary of any difficult terms
- a list of the command terms you will come across.

You will also see a list of other concepts that relate to the chapter. You should keep these in mind as you work.

Each related concept chapter is divided into topics that help you explore the concept through a variety of activities. Some activities can be done individually while others may be done with a partner or in a group. At the end of each topic is a reflection where you are given the opportunity to think about what you have learned and how it may relate to what you already know.

Here are some of the other features that you will come across in the book.

🌐 GLOBAL CONTEXTS

For many of the activities you will see an indication of a global context that is the focus of that activity. Global contexts help organize inquiry into six different areas.

- Identities and relationships
- Orientation in space and time
- Personal and cultural expression
- Scientific and technical innovation
- Globalization and sustainability
- Fairness and development

These global contexts indicate how the activity is relevant to your life and the real world.

ATL SKILLS

Alongside global contexts, each topic and activity includes an ATL skills focus. Usually, there are only one or two skills identified as the focus for an activity or topic. Of course, you will be using and developing other skills, but there is an emphasis on the particular skills in these boxes.

TIP

Throughout the chapters you will see additional information to help your understanding of a topic or activity.

QUICK THINK

These boxes provide questions to challenge your thinking. Your teacher may use them for a class discussion.

INTERDISCIPLINARY LINKS

As an MYP student you are encouraged to use skills and knowledge from different subject areas in your learning. Look out for these boxes which provide links to other subject groups.

CHAPTER LINKS

These boxes direct you to other chapters that relate to a topic or activity.

WEB LINKS

These boxes include websites and search terms for further reading and exploration.

1 Introduction to IB skills

Welcome to *Sciences: A Practical Guide* for MYP 4/5. This book will focus on some of the important skills in science—particularly those in scientific investigations and scientific inquiry.

This chapter gives a brief introduction to some of the essential elements of the MYP sciences.

Key concepts

In the Middle Years Programme (MYP), each subject area has key concepts that are used as a framework for knowledge. Key concepts are relevant within the subject and across disciplines and can help us make connections that transfer across time and culture. They are powerful ideas that we explore through different topics to try to understand the world around us. In MYP Sciences there are three key concepts that we use as the basis for study.

The three key concepts are:

- systems
- change
- relationships

Chapters 2 to 4 explore these key concepts.

Related concepts

Related concepts allow us to explore the key concepts in greater detail. They provide disciplinary depth and a focus for inquiry into subject-specific content.

There are eight related concepts that all three sciences (biology, chemistry and physics) share.

Related concepts shared by biology, chemistry and physics		
energy	evidence	models
interaction	form	movement
function	patterns	

Related concepts in science

Each of the three sciences—biology, chemistry and physics—also has four unique related concepts. Thus, each science has 12 related concepts: four specifically tailored to that science, and eight which are common to all.

In chapters 5 to 16, each shared related concept is dealt with in greater detail, as are four other related concepts (transformation, consequences, environment, balance) which are common to the modular sciences curriculum.

Through these related concepts, the book demonstrates and explains key skills and techniques in scientific investigation and scientific inquiry.

Learning skills

You know that you have been learning all your life. Firstly, you began in settings such as your home and neighbourhood. Then, your learning became more formal as you started school. Learning in the MYP is primarily inquiry-based learning. Your learning will continually cycle through three different phases.

Inquiry

Ask questions—it is the only way you are going to find out exactly what you want to know. This is especially true in sciences because asking the right question can lead to the right kind of experiment. Think about what you already know and what you want to know. Your curiosity is one of your best assets as a student.

Action

An important part of conceptual learning is action. Action in the MYP might involve learning by doing, service learning, educating yourself and educating others. Sometimes, you may choose not to act, based on newly acquired knowledge and understandings. Remember to think of the learner profile characteristic of being principled in your actions and make responsible choices.

Reflection

As a learner, you will become increasingly aware of the way that you use evidence, practise skills and make conclusions. Reflection on your learning helps you to look at the facts from a different perspective, to ask new questions and to reconsider your own conclusions. You may then decide to lead your inquiry in a different direction.

Inquiry learning can be frustrating. There is not always a "right" answer; sometimes conclusions may be uncomfortable or may conflict with what you want to believe, and you will come to realize that there are no "endpoints" in learning. As an MYP student, learning through inquiry, action and reflection is central to your education and forms the foundation of acquiring knowledge and conceptual understanding.

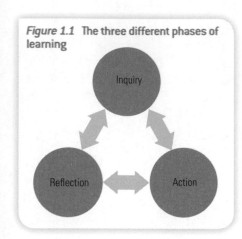

Figure 1.1 **The three different phases of learning**

Conceptual learning **is:**	Conceptual learning **is not:**
learning through inquiry	learning only through memorization
taking action to understand the world around you	trying to find the "right" answer
using knowledge to understand big ideas	passively accepting everything you read/hear/see.
making connections through concepts across different subjects.	

The characteristics of conceptual learning

The objectives covered

Both the skills and the objectives are closely related to the assessment criteria that your teacher will have made available to you. There are four assessment criteria and each one is designed to measure your skills in a different area of the sciences as follows:

Criterion A	Knowing and understanding	Maximum 8
Criterion B	Inquiring and designing	Maximum 8
Criterion C	Processing and evaluating	Maximum 8
Criterion D	Reflecting on the impacts of science	Maximum 8

The objectives of any MYP subject state the specific targets that are set for learning in the subject. They define what you will be able to accomplish as a result of studying the subject.

As mentioned, this book deals mostly with scientific investigation and scientific inquiry, and the criteria that focus on these skills are Criterion B and Criterion C. The objectives contained in these criteria represent the essential elements of the scientific method. In order to meet these objectives, you will engage in a variety of activities, continually refining your skills. These skills are very much interactive and interrelated, though in some instances your teacher may wish to deal with them as discrete skills.

Criterion B: Inquiring and designing

The objectives in Criterion B emphasize experimental work and scientific inquiry, particularly students' abilities to design a scientific investigation that allows them to collect sufficient data so that the problem or question can be answered.

These last two years of the MYP should prepare you to:

i) explain a problem or question to be tested by a scientific investigation

ii) formulate a testable hypothesis and explain it using scientific reasoning

iii) explain how to manipulate the variables, and explain how data will be collected

iv) design scientific investigations.

Criterion C: Processing and evaluating

The objectives in Criterion C focus on the collection, processing and interpretation of qualitative and quantitative data in order to reach appropriate conclusions.

These last two years of the MYP should prepare you to:

i) present collected and transformed data

ii) interpret data and explain results using scientific reasoning

iii) evaluate the validity of a hypothesis based on the outcome of the scientific investigation

iv) evaluate the validity of the method

v) explain improvements or extensions to the method.

Knowledge

The concept of "connections" or "perspective" is not something you can touch but you can certainly explain it to another person using specific examples from different subject areas. This is where your knowledge of facts is essential. Without the support of specific knowledge, facts and examples, it is very difficult to understand and explain key concepts and related concepts. In the MYP, your teachers have a choice as to what facts and examples they will use to help develop your understanding of key concepts.

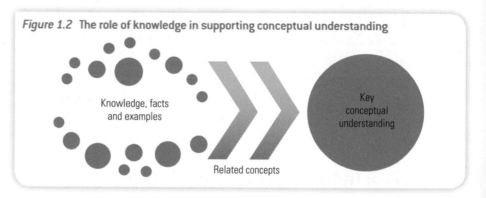

Figure 1.2 The role of knowledge in supporting conceptual understanding

Knowledge, facts and examples

Related concepts

Key conceptual understanding

The use of knowledge, facts and examples will be different in every MYP classroom but they will all lead you to an understanding of the key and related concepts in the sciences.

Global contexts

Now that you know what the key and related concepts are, let us focus a little more on the knowledge, facts and examples that will help you understand, explain and analyse them. The MYP calls this part of the curriculum global contexts. The global context is the setting or background for studying the key and related concepts. An easy way to think about global contexts is that they help your teacher choose the topics you are to study. There are six global contexts:

- identities and relationships
- orientation in time and space
- personal and cultural expression
- scientific and technical innovation
- globalization and sustainability
- fairness and development.

🌐 GLOBAL CONTEXTS

The choice of global context is influenced in several different ways.

Scale—study of a concept on an individual, local or global level.

Relevance—your education needs to be relevant for you and the world you live in, and this will influence the choice of context.

International-mindedness—IB programmes aim to develop internationally minded students and this is supported through using a variety of contexts to understand concepts.

Do students have an influence over what global context is chosen? Of course you do. It is the reason why the MYP sciences course looks different all around the world. The contexts that are relevant for you may not be relevant for a student studying in another country or even in another school in your own country. What all MYP sciences courses do have in common is the goal of deepening your understanding of the science key concepts.

Approaches to learning (ATL) skills

As a learner, you are developing a range of skills to help you learn and process significant amounts of knowledge and understanding. Some skills are very specific to particular subjects while other skills are ones that you use every day in every class, and will ultimately use for life! The skills that you learn through the MYP allow you to take responsibility for your own learning. There are five groups of ATL skills:

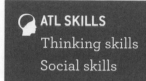

ATL SKILLS

Thinking skills

Social skills

Communication skills

Self-management skills

Research skills

Depending on the subject, you might focus more on one or two areas than on others. As you move through the MYP and mature as a student, the focus will also move through different skills—from being taught, to practising—to consolidate your skill ability. Read through the outline of ATL skills, taking some time to reflect on where and when you have learned, practised or mastered different skills. Also, think about which skills you still need to learn, practise or master.

Thinking skills	Critical thinking—the skill of analysing and evaluating issues and ideas.
	Creativity—the skill of exercising initiative to consider challenges and ideas in new and adapted ways.
	Transfer—the skill of learning by making new connections and applying skills, knowledge and understanding to new situations.
Social skills	Collaboration—the skill of working effectively with others.
Communication skills	Interaction—the skill of effectively exchanging thoughts, messages and information.
	Literacy—the skills of reading, writing and using language to communicate information appropriately and write in a range of contexts.
Self-management skills	Organization—the skill of effectively using time, resources and information.
	Affective skills—the skills of managing our emotions through cultivating a focused mind.
	Reflection—the skill of considering and reconsidering what is learned and experienced in order to support personal development through metacognition.
Research skills	Information and media literacy—the skill of interpreting and making informed judgments as users of information and media, as well as being a skillful creator and producer of information and media messages.
	Critical literacy—the skill of evaluating, questioning and challenging the attitudes, values and beliefs in written, visual, spoken and multimedia texts.

Approaches to learning (ATL) skills

It would be impossible to focus on all these areas in just your MYP sciences course in years 4 and 5, so we will be selecting specific skills to learn, practise and master in this book.

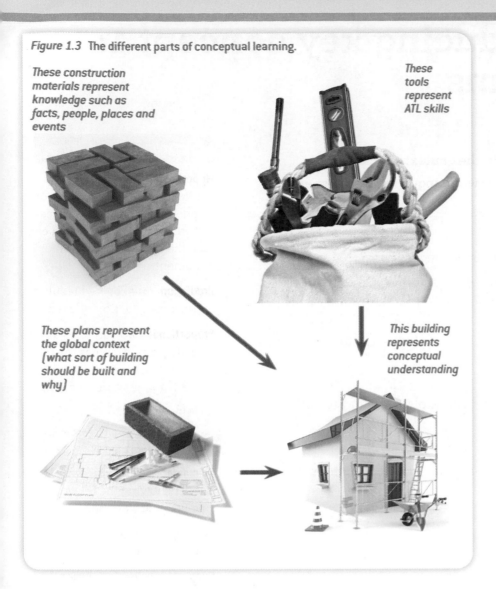

Figure 1.3 The different parts of conceptual learning.

These construction materials represent knowledge such as facts, people, places and events

These tools represent ATL skills

These plans represent the global context (what sort of building should be built and why)

This building represents conceptual understanding

Summary

Look back at Figure 1.3 on conceptual learning. Remember that conceptual learning happens when you use the inquiry cycle, develop your ATL skills and increase subject knowledge. These three factors work together to develop detailed understanding of the three key concepts in sciences: change, relationships and systems. While the content of sciences courses will look different in every MYP classroom, there is always the same focus on conceptual learning to construct a deeper understanding of the big ideas in life and the world around us.

Introducing key concept 1: systems

INQUIRY QUESTIONS	■ **What are systems in the context of science?**
	■ **What do we focus on when inquiring into systems in science?**

SKILLS	**ATL**
	✓ Make inferences and draw conclusions.
	✓ Analyse complex concepts and projects into their constituent parts and synthesize them to create new understanding.
	✓ Gather and organize relevant information to formulate an argument.
	Sciences
	✓ Formulate a hypothesis and explain it using scientific reasoning.
	✓ Solve problems set in unfamiliar situations.
	✓ Make connections between relevant information to draw conclusions.

GLOSSARY

Hypothesis a tentative explanation for an observation or phenomenon that requires experimental confirmation; can take the form of a question or a statement.

Prediction an expected result of an upcoming action or event.

Proportional two variables are directly proportional if the ratio of one to the other is always the same. If $x \propto y$, y/x is a constant.

COMMAND TERMS

Calculate obtain a numerical answer showing the relevant stages in the working.

Classify arrange or order by class or category.

Suggest propose a solution, hypothesis or other possible answer.

Introducing systems

Most people who have heard of Joel Salatin first read about him in Michael Pollan's best-selling book *The Omnivore's Dilemma,* or saw him in the documentary film *Food Inc.*

> *The shorter the chain between raw food and fork, the fresher it is and the more transparent the system is.*
>
> Joel Salatin

Considered by some a radical— and he describes himself as a "libertarian-environmentalist-capitalist-lunatic-farmer"—Joel is unique for his innovative and systems-based approach to food production.

He calls his 550-acre farm in Virginia, USA, Polyface Farm. There he raises cattle, chickens, pigs, turkeys and rabbits with the goal of "emotionally, economically and environmentally enhancing agriculture".

When questioned, he and the people who work on his farm say that they are actually grass farmers.

A system is a set of interacting or interdependent components

Everything is part of a system. For example, you are part of an ecosystem, and you are also part of the school system. In each of these systems, you interact with other parts of the system (such as the food you eat). Individual systems are usually part of bigger collections of many interdependent systems. For example, the ecosystem you are a part of interacts with the system by which carbon and carbon dioxide are cycled. And there is a whole ecosystem of bacteria in your gut.

Looking at things in terms of systems is based on the idea that everything is interconnected by a web of relationships. Parts of a system have common properties that work together. By studying parts of the system and how they interact, we can better understand how the system as a whole functions.

Systems and their surroundings can be as big as the planets and comets orbiting the Sun (the Solar System) or as small as the contents in a test tube in a chemistry lab.

It is chemistry that gives us one of the basic ways to describe systems— that is, whether a system is open, closed or isolated (Figure 2.1).

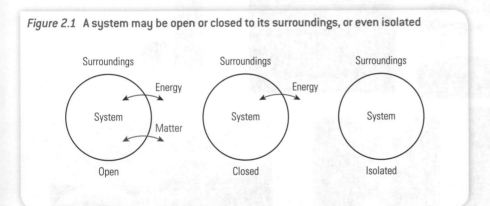

Figure 2.1 **A system may be open or closed to its surroundings, or even isolated**

Open systems are the most common in science. Necessary resources and materials are added to the system (called INPUTS) and can leave after being altered or transformed (called OUTPUTS).

In closed systems, materials cannot enter or leave the system, but energy can be exchanged with the surroundings.

An isolated system is one where neither matter nor energy can enter the system or leave it. This is a very unusual situation. Some people argue that only the Universe is an isolated system—but those who believe in multiple universes would disagree.

WEB LINKS

See what you can find out about grass farming and grass-fed livestock. You could search for "grass-fed cattle".

You will refer back to Joel's story as we think about systems in MYP Sciences.

CHAPTER LINKS

See Chapter 9 on interactions for more examples of interactions between two or more systems or bodies.

In small groups, consider the systems listed below. Try to **classify** them as open, closed or isolated. You will need to think about whether energy and/or matter can be exchanged with the surroundings.

Image	Type of system
Pot on stove with lid	
Pot on stove without lid	
Insulated bottle	
Earth	
Aquarium	
Biosphere II in Arizona	

🌐 **GLOBAL CONTEXTS**
Scientific and technical innovation

🧠 **ATL SKILLS**
Communication skills
Make inferences and draw conclusions.

When we consider what goes into and what comes out of individual systems, we start to appreciate how interconnected systems can be.

Sunlight is an output of the Solar System, and also an input for a meadow ecosystem. The meadow ecosystem has oxygen as an output. The oxygen is an essential input for the respiratory system of humans and all other animals in the meadow ecosystem.

Two of the principles of Joel Salatin's Polyface Farm are: (a) to have as few off-farm inputs as possible and (b) to have as few harmful outputs as possible. This makes the farm more sustainable (that is, able to meet the resource needs of all its components).

This contrasts strongly with the systems used to produce most of our food—systems generally called factory farming or industrial agriculture. Industrial agriculture has many off-farm inputs and many harmful outputs (Table 2.1).

An innovative way to control inputs and outputs in sustainable agriculture is to use something called an "eggmobile". Farmers like Joel Salatin recognize that cow manure is a valuable commodity that can be used to fertilize the grass pastures that cows feed on. At the same time, the cow manure can provide a valuable input in the food chain of egg-laying chickens.

When cattle move from one pasture to another, chickens in a mobile henhouse are moved on to the field after four days. During that four days, the eggs that flies have laid in the manure hatch into larvae. The chickens in the field scratch in the cow manure for the tasty larvae, which provide valuable nutrients. Because the cow manure has been scratched about by the chickens, it is broken up and dispersed through the grass. This means it fertilizes the grass pasture better than if it stayed in the clumps deposited by the cattle.

Reducing off-farm inputs and harmful outputs has the advantage of closing the system so energy and nutrients, and the processes that transform them, remain fixed and thus more sustainable. For example, the animals eat grass grown on the farm and their waste is used as a fertilizer. This reduces the need for additional feed and synthetic fertilizers, while the manure left by animals is not a problem to be removed.

Look at the advantages of pastured poultry for egg production. HD Karsten notes that eggs produced using this method have the following nutritional advantages over eggs from caged hens:

- two times as much vitamin E
- two times the total omega-3 fats (thought to help protect against heart disease)
- 40 per cent higher vitamin A concentration.

Inputs	Outputs
Fertilizer	runoff high in nitrates (eutrophication)
Pesticides	animal waste (pollution)
Antibiotics	pesticide effects on biodiversity
Grain/feed	topsoil (soil erosion)
Water/irrigation	water/runoff

Table 2.1 **A typical industrial agricultural production system is an open system with many inputs and outputs**

WEB LINKS
There is an entire movement dedicated to "pastured poultry" (poultry raised on grass pasture). Do some research and find out more about it. You might want to look at an organization called the American Pastured Poultry Producers' Association.

Many people consider Joel Salatin's approach to farming innovative, but it comes from a rich tradition of systems ecology.

The farm's name "Polyface" indicates Joel Salatin's deliberate introduction of a number of different animals into the agricultural system. This is to make the farm more like a natural ecosystem. More consumers in the food web makes better use of the grass as an energy resource, while avoiding the risks of diseases and pests that come from keeping large numbers of just one animal.

In Salatin's view, sustainable agriculture does not just make ecological sense, it also makes economic sense.

Systems provide structure and order

Human systems provide order and structure in built environments. Similarly, there are systems that provide order and structure in natural environments; for example, the ordered patterns observed in the night sky.

The clay tablet shown in Figure 2.2 dates back to around 1500 BCE in Babylonia. It outlines a series of observations about the "Bright Queen of the Sky", a reference to the planet Venus. This provides some of the earliest known evidence of astronomical observations.

From ancient times, the apparent motions of the stars has inspired and mystified observers. The ancient Greeks noticed wandering stars that did not behave like the others in the night sky. They named these wanderers *planates*. Eventually, astronomers worked out that there were a number of these planets making up our Solar System. But what caused the planets to move in the ways they did was a puzzle for many more centuries.

Two very different models resulted from the attempts to explain what was observed in the night sky. The earliest model was described almost 2,000 years ago by the astronomer Ptolemy. In this model, the Earth was at the centre of the Solar System. A second model was suggested by the astronomer Copernicus in the 16th century. In this later model, the Sun was at the centre of the Solar System.

Both Ptolemy and Copernicus were trying to predict the movement of planets in the night sky. Both made complex assumptions, many of which were incorrect. And neither model's **predictions** of the planets' movements were accurate.

Kepler's three laws of planetary motion gave a mathematical foundation to the model proposed by Copernicus with the Sun at the centre. Kepler produced equations for elliptical orbits of the planets round the Sun. These equations matched the data on the position and speed of the planets much better than circular orbits. Kepler's laws also eventually led Newton to his description of the law of universal gravitation.

Figure 2.2 **The Venus tablet of Ammisaduqa**

CHAPTER LINKS
See Chapter 4 on relationships for more information on Newton's law of universal gravitation.

The third of Kepler's Laws says that the time required for a planet to orbit around the Sun is **proportional** to the semi-major axis (the longest diameter) of its elliptical orbit.

What this means is that if you know how long it takes a planet to orbit the Sun, you can find out the dimensions of its orbit.

The relationship looks like this:

$$T^2 \propto a^3$$
$$\text{or } T^2 = 3 \times 10^{-4} a^3$$

where T is the period of the orbit in years, and a is the semi-major axis of the orbit in units of 10^{10} m.

For example, the Earth's orbit is 1 year and the semi-major axis is 15×10^{10} m, so substituting in the equation above:

$$1^2 / 15^3 = 0.000296 \text{ which is very close to } 3 \times 10^{-4}$$

Use this relationship to  the missing quantities for the following planets in our Solar System.

Planet	Semi-major axis (10^{10} m)	Period T (years)
Mercury		0.241
Venus	10.8	
Earth	15.0	1
Mars	22.8	
Jupiter		11.9
Saturn	143	
Uranus		84

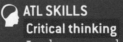 how satellite engineers could adjust the period of a communications satellite orbiting Earth so that it is exactly equal to the period of rotation of the Earth, 24 hours.

> **⊂⊃ LITERARY LINKS**
>
> Jonathan Swift, writing in *Gulliver's Travels*, knew all about Kepler's laws. The Laputa characters in the book:
>
> *... have likewise discovered two lesser Stars or Satellites, which revolve about Mars, whereof the innermost is distant from the Centre of the Primary Planet exactly three of his Diameters, and the outermost five; the former revolves in the Space of ten hours, and the latter in twenty-one and a half; so that the squares of their periodical times are very near in the same proportion with the cubes of their distance from the centre of Mars, which evidently shews them to be governed by the same law of gravitation, that influences the other heavenly bodies.*
>
> The periods and distances Swift gives are consistent with Kepler's laws (ie $T^2 \propto a^3$) for a satellite orbiting a planet.

> **🌐 GLOBAL CONTEXTS**
> Scientific and technical innovation

> **◉ ATL SKILLS**
> Critical thinking
> Analyse complex concepts and projects into their constituent parts and synthesize them to create new understanding.

CHAPTER LINKS
See Topic 1 in Chapter 16 on balance for more information on homeostasis in the body.

Systems can be static or dynamic

Some systems are static and unchanging. Other systems that are dynamic may appear to be stable and unchanging.

There are a number of systems in the human body that are balanced around an optimum value. Examples include the systems that maintain body temperature, water balance and blood sugar. There are constant changes and inputs to these systems, but a complex interaction of hormones, glands and nervous tissues monitors the body's internal conditions and works to maintain the optimum value for the system. This process is called homeostasis.

Activity 3   Thermoregulation

Your body maintains a constant temperature of around 37°C, whatever the external temperature or your level of activity. When the external temperature is below this point, which it usually is, the body must produce heat. This is accomplished during respiration by trillions of structures called mitochondria, which are found inside cells.

When the external temperature is extremely low or high, the body's systems respond in a number of ways to ensure that heat loss to the surroundings is minimized or excess heat is removed.

STEP 1   **Classify** each of the following as a response to low temperatures or high temperatures. Do some research and provide a brief explanation of how the mechanism of each response regulates body temperature (ie how the response alters how heat is lost or gained by the body).

Response	Temperature change	Mechanism
Shivering		
Sweating		
Goosebumps and erect body hair		
Reducing blood flow to extremities		

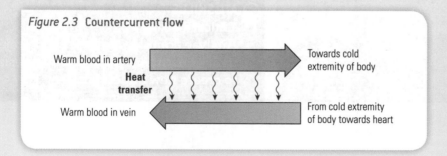

Figure 2.3 Countercurrent flow

Warm blood in artery — Towards cold extremity of body

Heat transfer

Warm blood in vein — From cold extremity of body towards heart

A further way the body conserves heat and avoids heat loss to the environment is called countercurrent flow. As shown in Figure 2.3, deep arteries and veins are very close to each other, but blood flows in opposite directions.

 **STEP 2** Examine the diagram in Figure 2.3. See if you can work out on your own how having relatively warm blood in the arteries from the heart moving past relatively cold blood returning from the hands and feet in veins would help to conserve heat.

STEP 3 Now pair up with a partner and share your ideas. See if you can put the ideas together to form a **hypothesis** about how this mechanism works.
Discuss your findings with your teacher if you are still unsure.

🌐 **GLOBAL CONTEXTS**
Scientific and technical innovation

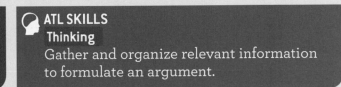 **ATL SKILLS**
Thinking
Gather and organize relevant information to formulate an argument.

Sustainable systems

A farmer in the 1970s would typically have grown more than one crop at different times. In any one area, farmers rotated crops (like wheat) with legumes (like soybeans) in order to restore nitrogen to the soil. Growing more than one crop at once provided some insurance against losses from extreme weather, and crop rotation reduced the build-up of pests, diseases and weeds in the soil. Although yields would have been relatively low, they would have been stable. The farming system ensured strong links between the agricultural production system and the natural ecosystem. The farm would have had few signs of degradation in soil or water quality—in other words, the farm would have been sustainable.

Modern agriculture bears little resemblance to these practices. It has a profit-driven approach that attempts to maximize yields at the expense of the ecosystem, and without minimizing off-farm inputs and harmful outputs. Herbicides and pesticides are used, and single crops (monocultures) are planted year after year. Monocultures remove nutrients from the soil so costly synthetic fertilizers are required to replace the nutrients. Monocultures are also extremely susceptible to change—including changes in weather, outbreaks of pests and weeds, and soil erosion.

The need for sustainable systems of food production will become a bigger issue in the future as the human population increases from around 7 billion toward what some scientists predict will be the eventual stable size of 9 to 10 billion people.

Summary

You have learned about the key concept of systems as it applies to science. You can now understand the importance of considering both the components of a system and how they interact. This will provide you with a perspective as you inquire further into the different kinds of systems in science, their structures, order and interrelatedness.

Introducing key concept 2: change

INQUIRY QUESTIONS	■ **What is change in the context of science?** ■ **What do we focus on when inquiring into change in science?**
SKILLS	**ATL** ✓ Listen actively to other perspectives and ideas. ✓ Inquire in different contexts to gain a different perspective. ✓ Practise visible thinking strategies and techniques. ✓ Collect and analyse data to identify solutions and make informed decisions. **Sciences** ✓ Interpret the relationship between two variables. ✓ Distinguish between correlation and cause and effect. ✓ Analyse information to draw justifiable conclusions. ✓ Make connections between relevant information to draw conclusions. ✓ Make connections between scientific research and related moral, ethical, social, economic, political or environmental factors.

GLOSSARY

Causal link a change in one variable causes a change in another.

Correlation a relationship between the two variables in an experiment. A correlation does not mean that a change in one variable causes the change in the other.

Scientific method the systematic way that scientists collect data through observation and experiment, to find the answers to scientific questions.

Variables quantities or characteristics of biological, chemical or physical systems that can have labels/names or numerical values.

COMMAND TERMS

Classify arrange or order by class or category.

Describe give a detailed account or picture of a situation, event, pattern or process.

Discuss offer a considered and balanced review that includes a range of arguments, factors or hypotheses. Opinions or conclusions should be presented clearly and supported by appropriate evidence.

Evaluate make an appraisal by weighing up the strengths and limitations.

Suggest propose a solution, hypothesis or other possible answer.

Introducing change

Ring around the rosy
A pocketful of posies
Ashes, ashes
We all fall down

> *There is nothing permanent except change.*
>
> Heraclitus

The nursery rhyme above may be familiar to you (perhaps with slightly different words) as it is often sung by young children in playgrounds and classrooms. The song might bring fun and laughter today, but the story behind it is much more serious.

WEB LINKS

Find out a bit more about the Black Death and how it first came to Europe. Your search should include "Messina" and "1347".

The line "ring around the rosy" is thought by some people to refer to the pimples that formed on those unfortunate enough to have contracted bubonic plague. This disease spread rapidly through Europe in the 14th century. We now call it the Black Death. It killed one in five people worldwide and, by some estimates, around half of Europe's population.

It took many centuries for the plague to be understood. There are still many regions of the world where the disease occurs, but one of the last major outbreaks in Europe happened in 1665. This was the Great Plague of London that killed about a fifth of its inhabitants. Those who had enough money tried to leave the city and find somewhere to stay in the countryside. There was less chance of coming into contact with the sick and dying in the rural environment.

When the plague appeared in Cambridge, the University of Cambridge closed the colleges and sent the students home. Among the university students returning to their villages and country towns was one of the greatest scientists and mathematicians of all time, a 23-year-old student named Isaac Newton.

Newton spent the next two years in his isolated country home. Those two years are among the most scientifically productive years of all time. Newton developed the foundations of modern calculus—the mathematics of change—and the foundations for his three laws of motion. These laws would explain why and how objects, even the planets, changed their motion.

You will come back to this story as you consider how to think about change in MYP Sciences.

Change alters an object or a system

The idea of change can describe a conversion or transformation from one behaviour, form or structure to another; in some cases, it could have a numerical value.

- In physics objects change their motion when acted on by an external, unbalanced force. Heating changes the temperature of an object.

- In chemistry substances change from one state of matter to another based on the motion of the particles within the substance. New substances can be made when elements or compounds react together.

- In biology species evolve as a result of the accumulation of many small genetic changes or mutations in a population over a long period of time. Your body and behaviour change during puberty, due to hormonal changes.

Activity 1 Physical and chemical changes

Change	Physical or chemical?	Reversible or irreversible?
Popping popcorn		
Melting butter for popcorn		
Exploding fireworks		
Rusting an old bicycle		
Frying an egg		
Breaking glass		
Burning paper		
Freezing ice cubes		

Questions

a) In small groups, **classify** each of the above changes into one of two categories—physical change or chemical change. Talk through your reasoning with each other.

b) In your groups, **discuss** whether each change is reversible or irreversible. **Describe** any pattern you can identify in the list.

 GLOBAL CONTEXTS
Scientific and technical innovation

 ATL SKILLS
Social
Listen actively to other perspectives and ideas.

For many centuries, one of the biggest questions about the plague was: Where is it coming from? The mystery continued until eventually the source was identified as a bacterium called *Yersinia pestis*. This bacterium is able to infect rats and humans. *Y. pestis* in the blood of infected rodents can be transferred by fleas that feed on the blood of both rodents and humans.

But there are not enough rats and fleas to explain how the plague bacterium *Y. pestis* suddenly became one of the most deadly organisms known to humans.

TIP

Rats carry fleas, which feed on their blood. If the rat is infected with the plague, some bacteria pass into the feeding flea. When that flea bites another rat or a human, it often regurgitates plague-infected blood into the wound. This infects the second rat or human with the plague

Change can be irreversible, reversible or self-perpetuating

Change is such a broad concept that we sometimes need to think about categorizing it into smaller and more useful units. In this way, we can consider different kinds of change.

Nowhere is this distinction more important than in our thinking about changes in the properties of elements and compounds.

A physical change is a change in which the substance changes form but keeps the same chemical composition.

A chemical change is a change in which the starting materials change into an entirely different substance or substances.

Physical and chemical changes can also be thought of as reversible or irreversible.

QUICK THINK

Some pathogens (organisms that cause disease) such as *Y. pestis* are able to survive and reproduce in both humans and other animals, but other pathogens cannot. What would have been different if *Y. pestis* were only able to survive in humans?

 Activity 2   Human changes to global cycles

Try to think of an example that shows a change from one form to another in at least one of the cycles listed:

- the rock cycle
- the water cycle
- the nitrogen cycle
- the carbon cycle.

Questions

a) **Describe** the change and what causes it.

b) **Suggest** one way that human activities are influencing this cycle.

GLOBAL CONTEXTS
Globalization and sustainability

ATL SKILLS
Transfer
Inquire in different contexts to gain a different perspective.

The plague and self-perpetuating change

Plague bacteria replicate rapidly in the bloodstream of an infected human or rodent, and the infection cycle is perpetuated or continued by fleas biting a new host. But flea bites cannot explain the rapid spread of the bubonic plague. The bacteria must be able to spread from person to person in more than one way. If spread of the disease was solely by flea bites, the entire history of the disease would probably have been very different.

So how were the bacteria transferred rapidly from person to person? Once the bacteria established themselves in the lungs of the victim, fever and severe coughing fits soon followed. When you cough, moisture droplets are spread into the air. When infected people began to cough up blood, the bacteria in the blood could pass via the moisture droplets to other people close by. This ability of the disease to self-perpetuate made the plague one of the deadliest killers in history.

Scientists consider causes and consequences of change

When scientists analyse observations or data, they look for a possible cause for the results. But the changes observed in an experiment may not be caused by the **variables** that were manipulated—a **correlation** between two variables does not necessarily mean that one causes the other (a **causal link**). There could be other factors involved in the process, or the correlation may be a coincidence.

It may seem pretty straightforward to control variables in the experiment so that other causes can be eliminated, but this is not always so. Natural phenomena are very complex, so establishing cause and effect is difficult and can sometimes lead to incorrect conclusions.

 Activity 3   **Analysing possible causes for differences in electricity usage**

How are human landscapes and resource use related?

In an analysis of the energy bills of 2 million homes in parts of the Western United States with a moderate climate, the energy supply company Opower found that homes with swimming pools use 49% more electricity and 19% more natural gas than those without.

STEP 1   Based on this evidence, write a conclusion about why energy consumption and having a pool are related. In other words, why do you think houses with swimming pools use 49% more electricity? The logical conclusion is that because swimming pools are very energy intensive (they need to be heated, pumped, filtered, and so on) they account for this significant difference in energy use.

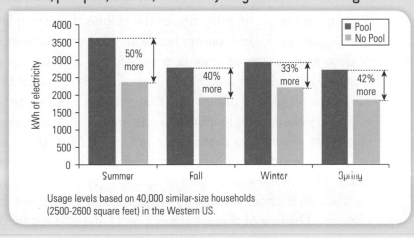

Usage levels based on 40,000 similar-size households (2500-2600 square feet) in the Western US.

- Effect: high electricity bills.
- Cause: heating and maintenance of swimming pools.
  But could there be some other factor or factors involved in this process?
The graph above, also using Opower's data but looking only at similar-sized houses, shows differences in electricity use in all four seasons.

STEP 2 Does this information reinforce the conclusion that pools account for the differences in electricity use in these houses? Why/Why not? Can you justify your answer?

If you thought the information made the case for a causal link weaker, you are correct. While there is no denying that pools use a significant amount of energy, there are a number of other factors that contribute to the 49% increase in electricity use for homes with pools.
These include:

- size of home—the average size of a home with a pool is 2,052 square feet (191 m²) vs 1,692 square feet (157 m²) for homes without pools
- number of occupants—homes with pools have a higher mean number of children (1.18 compared to the national mean of 1.08)
- income—the median income for houses with pools is $104,000 per year compared with the national median of $53,046.

STEP 3 Here is a useful visible thinking technique called "I used to think …; now I think …". Label the original conclusion you made about why pools with homes use more electricity "I used to think".

STEP 4 Now rewrite your conclusion based on the new information provided above. Label it "Now I think".

GLOBAL CONTEXTS
Orientation in space and time

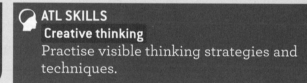
ATL SKILLS
Creative thinking
Practise visible thinking strategies and techniques.

Measure what is measurable, and make measurable what is not so.

Galileo Galilei

The mathematics of change

Remember that during the time of the plague, the young Isaac Newton left Cambridge for his country home. In the next two years, he was able to devote an enormous amount of time to some of the big questions that he was working on alone. This is particularly true of calculus. Newton struggled to understand how you could calculate how much a quantity changes in response to changes in another quantity—especially when those changes are not linear. After all, most processes in the real world do not have a linear relationship between cause and effect. The answer was elegant and simple. You need to break down the continuous changes into a very large number of infinitesimally small changes. The rate of change at any point can be found by finding the gradient of the straight line between the two points (Figure 3.1). For example, when a vehicle's velocity is changing, and the instantaneous acceleration (rate of change in velocity) is also changing over time, to calculate the acceleration, we need to calculate the change in velocity in an infinitesimally small time. In Figure 3.1, if a is very close to b, so that δx is very small, the average gradient between a and b ($\delta y / \delta x$) will be very nearly equal to the gradient at b.

This is complex, but the central idea is simple. And it has allowed us to analyse change in ways we could not have dreamed of without calculus.

We have the **scientific method** to explore the consequences of change in controlled, experimental conditions. A causal link between variables can only be verified after careful testing and critical review of the evidence by an international network of scientists.

The precautionary principle

In reality, there are many situations where the scientific evidence for a causal link is not yet complete. In these cases, it is uncertain whether the application of scientific innovations may have environmental impacts or consequences for human life or health. For example, research in the 1950s and 1960s led to a proven link between exposure to asbestos dust and a form of lung cancer called mesothelioma. However, throughout the period of research and even long after proof of the link was accepted, asbestos continued to be used in buildings. There was no action to reduce the exposure of workers to asbestos until the 1990s.

This has led to the creation of the precautionary principle, discussed by world leaders at the Earth Summit meeting in Rio de Janeiro in 1992. Most statements of the principle say that where an activity raises threats of harm to the environment or to human health, precautionary measures to prevent harm should be taken even if there is scientific uncertainty about cause and effect. This has implications for actions or policies by governments and businesses. Some people argue that the precautionary principle is not being met in the case of issues such as genetically modified foods, gene therapy, pollution, climate change, food safety, and others.

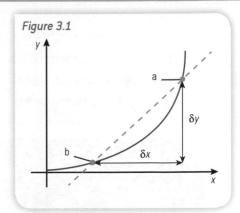

Figure 3.1

> ## INTERDISCIPLINARY LINKS
> Mathematics gives us the tools to measure rate of change using both calculus and algebra. Using graphs and graphing calculators, we can visualize change and make quick calculations about the instantaneous rate of change.

 Activity 4   **The precautionary principle**

Choose one of the following topics and research the problem the government was trying to solve, and the actions they took or policies they implemented to address the problem. Critically **evaluate** information and evidence from various sources and come to an informed opinion about these actions. Has this policy or action met the terms of the precautionary principle? What evidence did you find to help you reach your decision?

Topics

     a) Mad cow disease (BSE) and bans on animal feed

     b) Genetically modified crops that are resistant to pesticides

c) Use of DDT to control malaria

d) Nuclear power generation

e) Biological control agents

f) Blood transfusion and HIV transmission.

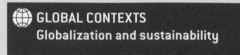
GLOBAL CONTEXTS
Globalization and sustainability

ATL SKILLS
Information literacy
Collect and analyse data to identify solutions and make informed decisions.

Cito, longe, tarde

When the plague arrived in Europe in the 14th century, doctors knew very little about the cause and prevention of the disease. They believed the illness was passed on through "corrupted air". As a result, much of the treatment involved perfumes and oils that, it was hoped, would clean the air and offer protection. You can see this in the original nursery rhyme, where the "pockets full of posies" (sweet-smelling herbs and flowers) were meant to protect children.

Plague doctors were special doctors who treated plague victims. They would often wear a beak-like mask (Figure 3.2) to protect them from the bad air thought to carry the disease.

The idea of corrupted or bad air resulted in the only sensible medical advice of the time: *cito, longe, tarde*. This Latin phrase translates as "flee quickly, stay long, return slowly". Changes in public health came about after the great plague. The idea of quarantine or isolating sick patients became standard and helped save countless lives during future outbreaks. There are also hints of future changes in the protective clothing of the plague doctor (Figure 3.2)—the use of gloves and special robes to protect clothes from contact with sick patients are part of modern hygienic practices. As for Newton, he heeded the advice—*cito, longe, tarde*—and avoided infection, and his three laws of motion allow us to both qualify and quantify changes in motion. His isolated study also led him to propose a solution to one of the greatest mysteries of his time—how and why the planets orbit the Sun. He called this invisible force of attraction between two bodies "gravity". Gravity revolutionized the study of astronomy, and also helped explain why a falling apple or other object accelerates to the ground.

Figure 3.2 A plague doctor. The long beak of the mask was stuffed with herbs and flowers meant to purify the air and protect the doctor from harm. Notice also the long robes and gloves to avoid contact with patients

Summary

You have explored the key concept of change as it applies to science in order to get a deeper understanding of how causal effects must be investigated carefully. This will provide you with a perspective as you inquire further into the different kinds of change in science, and their causes and consequences.

Introducing key concept 3: relationships

INQUIRY QUESTIONS	■ **What are relationships in the context of science?** ■ **What do we focus on when inquiring into relationships in science?**
SKILLS	**ATL** ✓ Collect and analyse data to identify solutions and make informed decisions. ✓ Interpret data. **Sciences** ✓ Process data and plot scatter graphs with a line of best fit to identify relationships between two variables. ✓ Analyse information to make a reasoned argument. ✓ Use appropriate scientific terminology to make the meaning of your findings clear.

GLOSSARY

Constant of proportionality in a directly proportional relationship, there is a non-zero constant with value k such that $y = kx$ for all values of x.

Inversely proportional two variables are inversely proportional if the product of the two is always the same. If $x \propto 1/y$, then xy is a constant.

Prediction an expected result of an upcoming action or event.

Proportional two variables are directly proportional if the ratio of one to the other is always the same. If $x \propto y$, then y/x is a constant.

Introducing relationships

On the night of 22 October 1701, almost 2000 sailors—among them two captains and the Commander-in-Chief of the British Navy, Sir Cloudesley Shovell (Figure 4.1)—were drowned after their ships were smashed against the rocks of the Isles of Scilly.

It remains one of the worst nautical disasters in Britain's history. When the government's inquiry into the tragedy was complete, the cause was no surprise.

Through centuries of ocean travel and commerce, many sailors and ships had been lost for the same reason—a ship's position in terms of longitude (how far east or west it was) was impossible to measure accurately.

> *The ability to relate and to connect, sometimes in odd and yet striking fashion, lies at the very heart of any creative use of the mind, no matter in what field or discipline.*
>
> Professor George J Seidel

COMMAND TERMS

Analyse break down in order to bring out the essential elements or structure. (To identify parts and relationships, and to interpret information to reach conclusions.)

Comment give a judgment based on a given statement or result of a calculation.

Discuss offer a considered and balanced review that includes a range of arguments, factors or hypotheses. Opinions or conclusions should be presented clearly and supported by appropriate evidence.

Figure 4.1 Admiral Sir Cloudesley Shovell

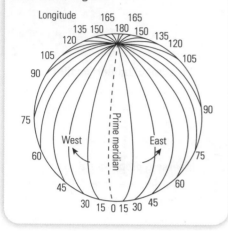

Figure 4.2 The prime meridian is a line of longitude, at which longitude is defined to be 0°. This meridian runs through Greenwich, England

WEB LINKS
Search for an online calculator that will find the distance of one degree of longitude at your latitude. Your search should include "length of a degree of latitude".

CHAPTER LINKS
Chapter 3 on change and Chapter 6 on evidence describe how scientists look for a possible causal link between variables.

At the same time, most maps and charts were inaccurate and incomplete. Without landmarks such as mountains to help navigation, a ship's captain could only estimate longitude by regularly noting the ship's speed and direction.

Measuring longitude requires knowledge of the shape of the Earth, the local time and the time, at the same moment, at some reference point. There is a mathematical relationship between time, latitude and distance. Because the Earth rotates on its axis once a day, we know that it goes through 360° of longitude in 24 hours. So, each hour of time difference between a ship and the port that it started from corresponds to 15° of longitude to the east or west. Because of the shape of the Earth, a degree of longitude at the equator where the circumference is biggest is a much longer distance than it is at the poles, where the distance is close to zero (Figure 4.2).

Nowadays, measuring time differences accurately is as easy as buying a cheap wristwatch. Set the watch to the correct local time when you leave port. While the watch continues to show the time at the port, you can measure the local time wherever you are by observing the angle of the Sun overhead. The time difference gives an extremely accurate measure of how many degrees of longitude have been travelled, using the relationship described above.

In an era of pendulum clocks, there was however no practical method for knowing the time at port while you were at sea. The early spring-based watches were not affected by the rolling of the ship but they were inaccurate because they were affected by changes in temperature. It meant that an error of even a few minutes over a month-long journey could quickly add up to being hundreds of miles off course. This could give tragic results, as for the British navy at the Isles of Scilly, or being lost at sea without sufficient supplies and a starving crew with scurvy.

The problem of longitude became a high priority for all seafaring nations. In 1714, the British government offered the Longitude Prize (£20 000—a sum equivalent to more than US$ 4 million today) to anyone who could solve the problem.

You'll come back to the problem of longitude as you think about relationships in MYP Sciences.

Relationships are connections

Determining relationships between variables gave scientists the ability to identify and understand connections. Some relationships are about shared properties (such as groups in the periodic table) or shared origins (for planets in the Solar System, or taxonomy), while others are connections between form or structure and function. Not all of these scientific relationships can be expressed quantitatively.

The relationship between the Earth and the celestial objects that move across the night sky was a question that occupied astronomers from ancient Maya and Greece, to the Ming dynasty and medieval Arabic and Persian observers. Once Kepler had established in the 16th century that the planets orbited the Sun in elliptical orbits, there was still the question of what caused these massive planets to repeat their journeys around the Sun over and over again.

The answer came ultimately from Isaac Newton, who realized that the force that causes objects to fall towards the Earth is the very same force that makes planets move around the Sun—the force he called gravity.

He proposed that there is a force of mutual attraction between any two masses—whether it is the Earth and an apple, or the Sun and Jupiter.

Newton showed that the strength of this attraction is directly **proportional** to the masses of the objects and is **inversely proportional** to the square of the distance between objects.

In mathematics, we can represent the idea of a proportional relationship using the symbol $\propto$.

So, for Newton's theory of gravitation the force of attraction F is proportional to the masses of the two objects, m_1 and m_2.

This relationship can be represented as: $F \propto m_1 \times m_2$

Similarly, the force of attraction is inversely proportional to the square of the distance d between the objects.

This relationship can be represented as: $F \propto 1 / d^2$

Putting all of this together and adding a **constant of proportionality**, the formula for the force of gravitational attraction is:

$$F = G m_1 m_2 / d^2$$

where G is the universal gravitation constant, the value of which was measured in 1798 by Henry Cavendish. In SI units G is equal to 6.67×10^{-11} Nm² / kg².

Simplifying relationships

Scientists often make approximations or assumptions to reduce complex algebraic formulae to simpler forms.

Calculating the gravitational force acting on objects is a very common calculation for astronomers studying stars and galaxies. As an example in a physics classroom, if we wanted to find out the gravitational force acting on a physics student with a mass 70 kg, we could use Newton's law of universal gravitation:

$$F = G m_1 m_2 / d^2$$

All of the quantities are known, or can be found—the mass of the Earth m_1 is 5.98×10^{24} kg. If the physics student is standing at sea level, he or she is a distance of 6.38×10^6 m from Earth's centre. For a large spherical object like the Earth, all its mass can be assumed to be concentrated at its centre. So the gravitational force is

$$F = 6.67 \times 10^{-11} \times (5.98 \times 10^{24}) \times 70 / (6.38 \times 10^6)^2$$

$$= 686 \text{ N}$$

However, using these variables in a complex formula every time you wanted to calculate an object's weight would quickly become tiresome. Some of the variables are always the same, such as for all objects on the surface of the Earth. So, is there a way to simplify this equation?

The mass of the Earth does not change, and neither does the gravitational constant, G. Your distance from the centre of the Earth may vary depending on your altitude, but altitude—even at the top of Mount Everest—does not make a significant difference compared to the distance from the centre of the Earth. So we can assume the distance d is the same for all objects on Earth.

Since Gm_{Earth} / d^2 is constant (does not change), we can write:

$$F = Gm_1m_2 / d^2 = mg$$

for a mass m at the surface of the Earth, where g is defined as gravitational field strength. (This will have a different value on a different body, such as the Moon).

The new equation, which you may be familiar with, gives the force of attraction of the Earth on a mass m. We call this weight.

⊂⊃ CHAPTER LINKS

In Chapter 10 on consequences you will read more about some of the consequences of climate change.

Change in a relationship brings consequences

When relationships change, some of the consequences may occur on a small scale, while others may be far reaching. Some may affect large networks and systems such as human societies and the planetary ecosystem.

You will read about global consequences in many of the chapters that follow, and this is particularly true of climate change. One of the problems of a warming planet is the increase in suitable conditions for mosquitoes. These insects thrive in regions with warm temperatures and high rainfall. Some mosquitoes spread malaria. As climate change continues, malaria may spread into new areas, although some areas may become malaria free.

No disease has affected more people than malaria. It is thought that as many as half of all people who have ever lived on Earth have been infected.

Activity 1   Prevention of malaria

Malaria is caused by a single-celled parasite of the genus *Plasmodium*. The parasite has a life cycle that requires both human and insect hosts. There is a complex relationship between the parasite, mosquitoes and humans. The parasite reproduces in the gut of mosquitoes. When the mosquito bites a human, the parasite passes into the person's bloodstream and travels to the liver. Here, the parasites mature and produce thousands more parasites, which then infect the host's blood leading to illness and often death. Mosquitoes take in the parasite when they feed on an infected person, and so the cycle continues.

Do some research in groups and analyse the main methods currently being used to control the spread of malaria:

- preventing the breeding of mosquitoes
- preventing mosquitoes from visiting/entering houses
- preventing mosquitoes from biting people
- killing mosquitoes by spraying insecticides
- treating people with malaria (this may include testing).

Make a recommendation as to which method you think would provide the greatest impact for a proposed investment, particularly for saving the lives of children. Justify your choice.

Some things your group should discuss and consider when justifying your recommended method:

a) Which strategies are the most effective in preventing the spread of malaria?
b) How much is each plan likely to cost?
c) How many lives are your recommendations likely to save?
d) Who should pay for these measures? Suggest some ways that reduced malaria transmission will have economic benefits.
e) Discuss the statement: "Why should anyone care about things that happen in other countries?"

🌐 **GLOBAL CONTEXTS**
Fairness and development

🧠 **ATL SKILLS**
Information literacy
Collect and analyse data to identify solutions and make informed decisions.

Determining the ripeness of grapes and making the decision when to harvest is extremely important to wine-makers. It has an enormous impact on the quality of the fruit juice or wine that can be made, and on alcohol content that will result if the juice is fermented. Wine-makers use an instrument called a refractometer to indirectly measure the sugar content of grape juice and make decisions about harvesting (Figure 4.3).

In many wine-producing nations, the sugar content is given a special unit called degrees Brix (symbol °Bx). One degree Brix is equivalent to 1 gram of sucrose in 100 grams of solution.

Figure 4.3 Refractometer for measuring the ripeness of grapes. The same method is used by optometrists to measure the refractivity of a patient's eyes

In California, one of the world's largest and most important wine-producing areas, there is evidence that sugar content in grapes has been increasing as a result of global warming.

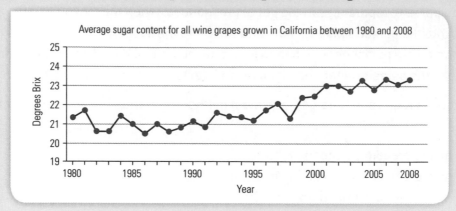

Average sugar content for all wine grapes grown in California between 1980 and 2008

STEP 1 Use the data above to quantify the change in °Bx over time. For this you will need to draw a line of best fit and then find the slope of the line. The unit will be °Bx/year.

STEP 2 Based on your calculation, would you say that there is a large increase?

For wine-makers, the sugar content of the grape juice is extremely important because this is the raw material that gets converted to alcohol during the fermentation process. Sweeter grapes make wines with greater alcohol contents.

While this may not sound like such a bad thing, in fact, alcohol content in wine is meant to be at a certain concentration—typically 12–14 per cent by volume.

Potential alcohol content as percentage alcohol by volume (%abv) is calculated using the formula: $\%abv = °Bx \times 0.59$

So one degree Brix would yield an alcohol content of 0.59 per cent, while 30 °Bx would yield an alcohol content of 17.7 per cent.

STEP 3 Use the data from the graph to determine the potential alcohol content from grapes grown in 1980 and grapes grown in 2008.

STEP 4 After calculating these values, **comment** on your answer to the question about whether or not the sugar content shows a large increase.

STEP 5 **Discuss** with a partner the implications you think climate change might continue to have on the sugar content in grapes. Suggest how wine-makers might deal with this problem in the future.

GLOBAL CONTEXTS
Scientific and technical innovation

ATL SKILLS
Critical thinking
Interpret data.

Relationships are found through observation or experimentation

Long before a prize was advertised for solving the longitude problem for navigation at sea, a number of renowned scientists and astronomers had been working on an alternative method that did not require an accurate clock to show the reference time. The method used a **prediction** of the Moon's apparent motion relative to the stars as a clock to find a reference time.

From around 1660, observatories had been established in major cities around the world to study the Moon and stars and to painstakingly log their movements and relative positions to one another at precise times of day at precise times of year.

The result was a growing body of knowledge in the form of very detailed charts of the Moon's future positions. The only problem was trying to recreate the kind of careful measurements required to use the lunar charts while standing on a pitching and rolling ship in the middle of the ocean.

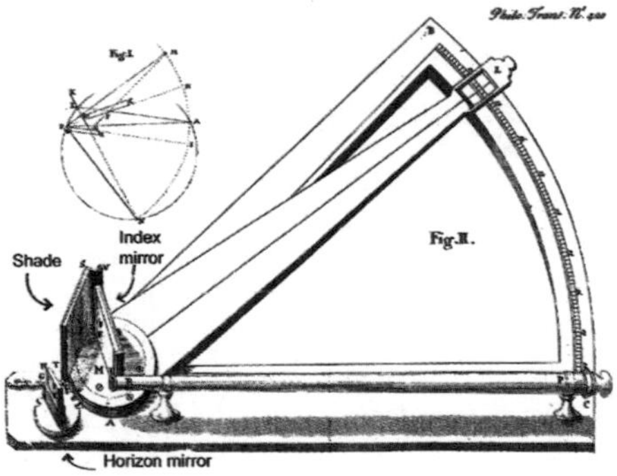

This problem was mostly solved with the invention of an amazing device called an octant (Figure 4.4). It allowed sailors to measure angular distances between the Moon and other celestial bodies. Suppose, for example, a navigator was able to determine that the Moon was 25 degrees away from Sirius—the so-called Dog Star—at a time of 1:00 am onboard the ship. The star chart might call for this configuration at 3:00 am in London. This would mean the ship's time was two hours behind London and that the ship was therefore 30° west of London.

This method came close to solving the longitude problem, but it required clear skies and someone with a very strong understanding and experience of mathematics and astronomy, conditions that were frequently not fulfilled at sea.

One of the consequences of not being able to find longitude at sea was not remaining on course. This meant already long journeys could become even longer. Ships had to carry enough food for the entire crew until the next port. This was difficult to plan even at the best of times. As a result, it was the sailors who suffered when there were navigational errors. More often than not, the lack of fresh food meant the crews suffered hunger and frequently also scurvy—a serious and painful disease we now know is caused by a lack of vitamin C. This vitamin is found in fresh fruits and vegetables but during long sea voyages sailors were unlikely to have either.

If left untreated, scurvy can kill. In 1553, Admiral Sir Richard Hawkins observed that during his 20-year career, 10,000 men under his command in the British Navy had died of scurvy. A way of preventing the disease had yet to be discovered.

Figure 4.4 John Hadley invented an octant with mirrors for measuring the altitude of the Sun

I wish some learned man would write of it, for it is the plague of the sea, and the spoyle of mariners.

Sir Richard Hawkins

It was not until 1747 that James Lind, a ship's doctor on HMS (His Majesty's Ship) *Salisbury*, conducted a small controlled experiment on sailors with the disease. Twelve patients were given a similar diet but six pairs were each given a different daily treatment (including vinegar, seawater and cider). After six days, only the pair who were given two oranges and one lemon per day showed any signs of recovery.

Sadly, despite the link between citrus fruit and the treatment of scurvy, it would be years before citrus fruit became an obligatory part of the diet onboard ships. But when it did, it meant the end of this terrible disease for sailors.

The problem of the accurate determination of time on board ship and on shore, so that the ship's longitude could be calculated, was eventually solved by a brilliant, self-taught watch-maker from England. John Harrison created a watch so accurate that, when it was tested in 1761 on a transatlantic crossing from England to Jamaica, it lost only 5 seconds during the 81-day voyage. In comparison, modern clocks and wristwatches with quartz movements are accurate to within 5–20 seconds per year.

Harrison understood that even small changes in temperature that happen day to day and when ships move from northern to southern latitudes could thin or thicken the lubricating oil in most watches. This made their inner metal workings expand or contract. Changes in barometric pressure also affected the metal workings. But Harrison succeeded in solving these challenges where all others had failed. After a lifetime of trying to solve the problem of longitude and several different versions of his watch, also called a chronometer, John Harrison was finally awarded the Longitude Prize in 1773.

Summary

You have looked at the key concept of relationships in this chapter in order to get a deeper understanding of how important and how prevalent it is in science, and how relationships can be expressed quantitatively or by the grouping of qualitative characteristics. You will be able to identify examples of when changes in relationships lead to consequences, and begin to search out the connections between variables.

Energy

INQUIRY QUESTIONS

TOPIC 1 Energy content of diets

- **How can the energy content of food be determined?**
- **What factors influence the energy content of food?**

TOPIC 2 Energy changes in chemistry

- **How can the energy change for a phase change or chemical reaction be measured?**
- **What factors influence the temperature change in a reaction?**

TOPIC 3 Sustainable energy resources

- **What is the most sustainable way to produce energy worldwide?**
- **How can the efficiency of wind turbines be improved by designing and testing new models of blades?**

SKILLS

ATL

✓ Evaluate evidence and arguments.

✓ Collect and analyse data to identify solutions and make informed decisions.

✓ Organize and depict information logically.

✓ Draw reasonable conclusions and generalizations.

✓ Use and interpret a range of discipline-specific terms and symbols.

✓ Test generalizations and conclusions.

✓ Interpret data.

✓ Use models and simulations to explore complex systems and issues.

Sciences

✓ Formulate a testable hypothesis.

✓ Design a method for testing a hypothesis, and select appropriate materials and equipment.

✓ Explain how to manipulate variables, and how enough data will be collected.

✓ Organize and present data in tables ready for processing.

✓ Interpret data gained from scientific investigations and explain the results using scientific reasoning.

✓ Evaluate the validity of a model based on the outcome of an investigation.

✓ Evaluate the validity of the method.

GLOSSARY

Control variables the variables that remain constant and unchanged in an experiment, to investigate the effect of changing the independent variable.

Dependent variable the variable in which values are measured in the experiment.

Extensions to the method developments for further inquiry as related to the outcome of the investigation.

Independent variable the variable that is selected and manipulated by the investigator in an experiment.

COMMAND TERMS

Design produce a plan, simulation or model.

Evaluate make an appraisal by weighing up the strengths and limitations.

Formulate express precisely and systematically the relevant concept(s) or argument(s).

Measure obtain a value for a quantity.

✓ Describe improvements to a method, to reduce sources of error, and possible extensions to the method for further enquiry.

OTHER RELATED CONCEPTS   **Balance**   **Consequences**   **Transformation**

Introducing energy

It is not by chance that energy is the first related concept in this book. Energy and the law of conservation of energy are fundamental to all disciplines in science. Its study and description can seem elegant and simple—energy can be transformed but not created or destroyed and it is available in unlimited amounts from the Sun. Yet as a concept, it can be difficult to understand.

What exactly is energy? In the MYP, we define energy as the capacity of an object to do work or transfer heat. Work is done when a force moves, and the unit of work is joules (J), the same as the unit of energy. This may instantly make sense to some of you but it may leave others wondering what that means exactly.

Scientists and entrepreneurs are constantly on the lookout for new energy technologies because humans want vast amounts of energy. Ideally, we would like our energy supplies to be sustainable. But most energy resources cannot be used without environmental consequences.

The World Energy Council refers to energy sustainability in terms of three core dimensions—environmental sustainability, energy equity, and energy security—calling it the energy trilemma. Everyone wants their energy supply to be environmentally friendly, cheap and reliable (available at all times). This is probably impossible. What usually happens is that two of the three criteria are pursued at the expense of the third.

In this chapter, you will look at ways in which energy is used and at environmentally sustainable ways of producing energy. You will also consider the cost and reliability of these energy resources.

One of the energy resources that holds great promise as a clean and inexhaustible energy source for the future is also one of the most fundamental in the Universe. It is described by perhaps the most famous scientific equation, $E = mc^2$.

This resource is behind the energy source that rises in the sky every morning to light our world and power the Earth's food chains through photosynthesis by green plants. Yet, despite decades of research, scientists and engineers are still no closer to commercializing this energy resource (Figure 5.1).

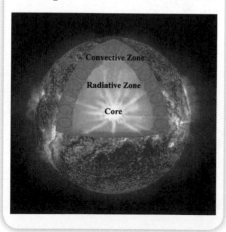

Figure 5.1 Fusion reactions in the core of the Sun release energy. By the time the energy reaches the surface of the Sun, it is mostly in the form of infrared radiation, visible light and some ultraviolet radiation

The process is called nuclear fusion. In fusion, the nuclei of small atoms collide at great speeds and fuse to form a larger nucleus of a new atom. The product of the collision has slightly less mass than the original two reactants, and this apparent loss of mass (the m in the equation $E = mc^2$) is released as energy. Because c, the speed of light (3.0×10^8 m/s), is so large, even a tiny change of mass will result in a huge amount of energy being released.

Consider the fusion of one atom of deuterium with one atom of tritium (both isotopes of hydrogen with one proton but differing numbers of neutrons) to form one atom of helium and a neutron. This is the most promising method of nuclear fusion being investigated at several research reactors.

$$_1^3H + {}_1^2H \rightarrow {}_2^4He + {}_0^1n$$

	$_1^3H$	$_1^2H$	$_2^4He$	$_0^1N$
mass ($\times 10^{-27}$ kg)	5.007	3.343	6.645	1.675

Total mass before the reaction $= (5.007 + 3.343) \times 10^{-27} = 8.350 \times 10^{-27}$ kg

Total mass after the reaction $= (6.645 + 1.675) \times 10^{-27} = 8.320 \times 10^{-27}$ kg

Mass defect $= (8.350 - 8.320) \times 10^{-27} = 0.030 \times 10^{-27}$ kg

Energy released:

$$E = mc^2$$
$$= 0.030 \times 10^{-27} \times (3 \times 10^8)^2$$
$$= 0.27 \times 10^{-11} \text{ J}$$

This may seem like a tiny amount of energy, but 1 kg of tritium contains 1.99×10^{26} atoms, giving a total energy released per kilogram of $0.27 \times 10^{-11} \times 1.99 \times 10^{26} = 5.37 \times 10^{14}$ J.

Compare this to the energy released from burning one litre of gasoline, which is 3.2×10^7 J.

One of the raw materials for fusion, deuterium, comes from seawater. There are enough deuterium atoms in the world's oceans to meet current global energy needs for billions of years—about as long as our Sun is expected to last—with very few environmental drawbacks, since fusion produces only helium as a by-product.

So why are we not all using fusion power to replace fossil fuels and nuclear power, and to solve our energy needs? Is it because one of the three aspects of the energy trilemma is not being met? Or are there technical barriers to being able to meet current and future demand?

These are the kinds of open questions you will be challenged with throughout this book. Through research, inquiry, experimentation, discussion and exchange of ideas you'll be exposed to concepts in a way that will help you to see the connections between these ideas. You will also be able to use your skills to tackle life beyond the classroom.

TOPIC 1

Energy content of diets

The chemical energy stored in food is converted by the mitochondria in our cells into a usable form in order to provide energy for chemical reactions.

Not all food contains the same amount of energy per unit of mass and not all food is equally nutritious. Nutrition information panels on the side of food packets help consumers make informed choices. The first box usually gives information about the energy content of the food.

Calories and joules are both units of energy. You will hear calories mentioned most often in the popular media, but scientists use the SI unit, the joule (J). The units of energy in food nutrition panels vary from country to country, though the metric standard is to indicate the energy content in joules. One calorie is approximately 4.2 J but Calories written with an upper case C means 1,000 calories or 1 kilocalorie.

An average adult should consume a minimum of 2,000 kilocalories per day or about 8,400 kilojoules (kJ).

Measuring energy changes

The energy content of food can be determined experimentally using a procedure called calorimetry. If a sample of known mass is burned, the energy released can be used to warm a known mass of water.

The specific heat capacity of a substance is the amount of energy it takes to warm 1 g of the substance by 1°C. The specific heat capacity of water is 4.18 J/g K so it takes 4.18 J of energy to raise the temperature of 1 g of water by 1°C .

The calorimetry apparatus consists of a thermometer to detect a change in temperature and a container with a known mass of water. The sample to be tested is ignited and placed below the container with the water and the temperature change of the water is recorded (Figure 5.2).

The formula used to measure the energy transferred to the water by the burning food, and so determine the energy content, is:

energy transferred (in J)

= mass of water (in g) × specific heat capacity of water (in J/g K) × change in temperature (in K)

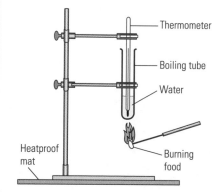

Figure 5.2 Simple apparatus to measure the energy content of food

- Thermometer
- Boiling tube
- Water
- Burning food
- Heatproof mat

Activity 1 — Measuring the energy content of different foods

You are going to investigate how different types of food preparation or cooking affect energy content. For example, roasted peanuts could be compared to fresh peanuts taken out of their shells. Alternatively, you could measure the energy content of foods such as potato chips, mini marshmallows, bread or pasta.

STEP 1 Gather together the following materials and apparatus:

- eye protection
- clamp and stand
- thermometer (0–110°C)
- heatproof mat
- boiling tube
- measuring cylinders (10 or 50 cm³)
- top pan balance
- various foods to test
- mounted needle
- wooden splints for lighting the food.

STEP 2 Read through the procedure and construct a data table to record the data you are going to collect.

STEP 3 Put 10 cm³ of water in a boiling tube. Clamp the boiling tube in the retort stand as shown in Figure 5.2 above.

STEP 4 **Measure** the mass of a small piece of food in grams and record it in the table.

STEP 5 **Measure** the temperature of the water in the test tube and record it in the table.

STEP 6 Fix the food on the end of the mounted needle. If the food is likely to melt when heated put it on a teaspoon instead of on the needle.

STEP 7 Light the food using a wooden splint. As soon as the food is alight, hold it about 1 cm below the boiling tube. If the flame goes out, quickly relight it.

STEP 8 When the food stops burning, stir the water in the tube with the thermometer. **Measure** and record the highest temperature reached.

> **TIP**
> If you do not get a significant temperature rise, repeat using a larger portion size, such as five potato chips.

STEP 9 Empty the boiling tube and refill it with another 10 cm³ of water. Repeat the experiment using a different food each time.

> **TIP**
> When you calculate the energy transferred, make sure you use units consistently. If the heat capacity is given per kilogram of water, you should convert a mass in grams to a mass in kilograms.

STEP 10 Calculate the temperature change for each food you tested. Use this to calculate the energy transferred to the water for each food you tested. Remember, since the density of water is 1 g/cm³, 1 cm³ of water has a mass of 1 g.

STEP 11 Calculate the energy transferred per gram of food for each sample.

STEP 12 Which food has the highest energy value?

🔗 LITERARY LINKS
Richard Wrangham, *Catching Fire: How Cooking Made Us Human*. Basic Books, New York, 2009.

In this book, the author explains the hypothesis that because cooking reduces the energy required to digest food, the boost in available energy of cooked food must have resulted in major transformations in human cultural and biological evolution.

🌐 GLOBAL CONTEXTS
Scientific and technical innovation

🧠 ATL SKILLS
Critical thinking
Evaluate evidence and arguments.

 Activity 2   Monitoring dietary energy intake

Based on information contained in databases about the composition of the foods consumed, the energy intake of an individual can be tracked and compared to the recommended intake.

Food choices have an impact on human health. A healthy lifestyle involves being informed about dietary intake and dietary energy requirements. By collecting and analysing data from a diet tracker, you should be in a position to make informed decisions about dietary energy intake.

STEP 1 Over the period of one week, enter your food consumption into the "Super Tracker" app available from the US Department of Agriculture (USDA) at www.supertracker.usda.gov.

STEP 2 Collect data for the whole class and transform the data into a suitable chart. For example, you could calculate mean daily energy intake and compare this to recommended daily intake. If you are not comfortable sharing this information with classmates, talk to your teacher and make another arrangement.

🔗 INTERDISCIPLINARY LINKS
The concepts of "choices", "balance" and "energy" also appear in the MYP Physical and Health Education (PHE) course. See if any of the data you collected could be used in your PHE classes.

🌐 GLOBAL CONTEXTS
Scientific and technical innovation

🧠 ATL SKILLS
Information literacy
Collect and analyse data to identify solutions and make informed decisions.

Dietary energy from saturated fat

Not all forms of dietary energy have the same health impact. Scientists have noted that the incidence of coronary heart disease varies significantly across cultures, and that cultures differ in their diets. The Seven Countries Study was carried out from 1958 to relate diets rich in saturated fat to the incidence of deaths in middle-aged men due to coronary heart disease. The data is shown in Table 5.1.

Mortality rate indicates the number of people who die out of some number of the population over a certain amount of time. Thus, a mortality rate of 500/100,000/year means that out of 100,000 people in the population, 500 die every year of that particular cause.

Region	% Calories as saturated fat	Coronary heart disease mortality rate /100000/y	Mortality rate due to all causes /100000/y
East Finland	22	992	1727
West Finland	19	351	1318
Zutphen (Netherlands)	19	420	1175
USA	18	574	1088
Slavonia (Croatia)	14	214	1477
Belgrade	12	288	509
Crevalcor (Italy)	10	248	1241
Zrenjanin (Serbia)	10	152	1101
Dalmatia (Croatia)	9	86	758
Crete	9	9	543
Montegiorgio (Italy)	9	150	1080
Velika (Croatia)	9	80	1078
Rome	8	290	1027
Corfu	7	144	764
Ushibuka (Japan)	3	66	1248
Tanushimaru (Japan)	3	88	1006

Table 5.1 Mortality rates and percentage of calories as saturated fat in several countries

 Activity 3   Analysing dietary energy from saturated fat

In this activity, you will construct a graph to visually compare the data from a number of European regions as well as the USA.

STEP 1 Open a spreadsheet program.

STEP 2 Enter the data from Table 5.1 above.

STEP 3 Use your spreadsheet program to create a scatter graph of mortality rate (all causes) and mortality rate (coronary heart disease) against the percentage of calories as saturated fat.

STEP 4 Add a y-axis title "mortality rate". Include the units.

STEP 5 Add an x-axis title "percentage calories saturated fat". Include the units.

STEP 6 Add a trendline for each data series.

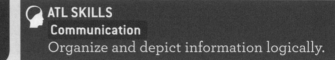
WEB LINKS
Have a look at how to add a trendline using Excel. Go to www.youtube.com and search for trendline Excel.

🌐 GLOBAL CONTEXTS
Scientific and technical innovation

ATL SKILLS
Communication
Organize and depict information logically.

LITERARY LINKS
In his book *In Defense of Food*, Michael Pollan decries "nutritionism", which is the tendency to view the food that we eat as a collection of constituent nutrients to be avoided or increased in our diet.

Reflection on Topic 1

A very similar reductionist approach surrounding energy or caloric intake is the basis of many of the weight-loss products, programmes and books that make up a multibillion-dollar industry.

Consider Pollan's basic food motto: "Eat food, not too much, mostly plants."

How does this less complicated and holistic approach to eating food match with what you have learned in this topic?

TOPIC 2

Energy changes in chemistry

Changes in states of matter either require energy (for example, melting of solid ice to liquid water) or release energy (for example, condensing of steam to liquid water). The law of conservation of energy states that in a closed system, energy can neither be created nor destroyed. So when we say energy is required, it must be supplied from elsewhere in the system, which loses the same amount of energy. Similarly, if a reaction releases energy, this energy does not disappear—it is absorbed by another part of the system.

A chemical reaction involves breaking of bonds in the reactants and forming of bonds in the products. It takes energy to break bonds (work is done on the reactants) and when bonds form, energy is released.

If more energy is released when the bonds of the products are formed than it takes to break the bonds of the reactants, the reaction is said to be exothermic (energy is released to the surroundings). The total energy of the system remains the same, but the reactants had more stored energy than the products.

A graphical way of representing this is shown in Figure 5.3.

If it takes more energy to break the bonds of the reactants than is released by the formation of the products, the reaction is said to be endothermic (energy is absorbed from the surroundings). The total energy of the system of reacting chemicals and surroundings remains the same, but the products have more stored energy than the reactants.

A graphical way of representing this is shown in Figure 5.4.

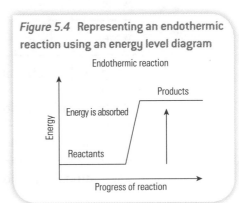

Figure 5.3 Representing an exothermic reaction using an energy level diagram

Figure 5.4 Representing an endothermic reaction using an energy level diagram

The relationship between energy changes and temperature change

When energy is released in an exothermic reaction, the temperature of the surroundings (the reaction mixture) increases. When energy is absorbed in an endothermic reaction, the temperature of the surroundings decreases. Using a thermometer or temperature probe to measure temperature changes shows whether chemical reactions are exothermic or endothermic.

 Activity 4   **Endothermic and exothermic changes**

In this activity, you will determine whether changes are exothermic or endothermic based on observed temperature changes.

STEP 1   Gather together the following materials and apparatus:

- eye protection
- thermometer (−10°C to 110°C)
- boiling or test tubes
- test tube rack
- graduated cylinder (10 cm³)
- spatula
- glass stirring rod.

(SAFETY) Dilute hydrochloric acid is an irritant: avoid contact with eyes, mouth and skin. Wear eye protection at all times. Copper(II) sulfate is harmful if swallowed and irritates the eyes and skin. Calcium chloride is an irritant. Sodium hydroxide is corrosive.
Your teacher will give you the liquids and solids you are going to use.

STEP 2   **Design** a data table to write down data that are to be collected, and how the data will be processed and classified.

STEP 3 For each row in the table below, add 5.0 cm³ of the liquid to a boiling tube, record the initial temperature of the liquid and add one spatula of the solid.

Liquid	Solid
water	sodium nitrate
water	ammonium chloride
water	barium chloride octahydrate
water	calcium chloride
copper(II) sulfate solution (1M)	zinc metal
2M hydrochloric acid	sodium hydroxide
2M hydrochloric acid	calcium carbonate
2M hydrochloric acid	3 cm length magnesium ribbon

STEP 4 Stir well and record the final temperature.

a) In which experiment was heat energy given out?

b) Which experiment was the most exothermic? Explain how your data shows this.

c) For one of these experiments, suggest how a greater temperature change could be produced.

GLOBAL CONTEXTS
Scientific and technical innovation

ATL SKILLS
Critical thinking
Draw reasonable conclusions and generalizations.

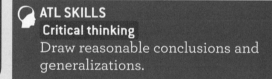

WEB LINKS
A summary of the terms exothermic and endothermic can be found by going to: www.youtube.com and searching for Tricky Question: Exothermic or Endothermic.

Calculating heat transferred

The heat transferred from one substance to another can be calculated by measuring a change in temperature. The amount of heat energy, measured in joules (J), required to raise the temperature of one gram of a substance by one degree Celsius is called the specific heat capacity. The specific heat capacity of water is 4.18 J per gram per degree Celsius and has the symbol c.

The amount of heat transferred, Q, can be calculated by multiplying the mass of water, m, by specific heat capacity c and the temperature change ΔT:

$$Q = mc\Delta T$$

Activity 5 — Energy transfer between hot and cold water

When hot and cold water are mixed in an insulated container, what will be the final temperature of the mixture? You will need to calculate the heat energy lost by the hot water, and the heat energy gained by the cold water. In order to calculate the energy transfer, you will need to ensure that you have recorded the initial and final temperatures.

Procedure A
Read through the procedure and construct a data table to record the data you are going to collect.

Procedure B
Repeat as in procedure A, but this time add 50 cm³ of hot water to 100 cm³ of cold water.

Procedure C
Repeat as in procedure A, but this time add 50 cm³ of cold water to 100 cm³ of hot water.

STEP 1 Gather together the following materials and apparatus:

- an insulated container such as a camping mug or polystyrene foam cup
- graduated cylinder (50 cm³)
- thermometer (0–110°C).

STEP 2 **Measure** 50 cm³ of cold water and transfer it to the insulated cup. Record the initial temperature of the cold water.

STEP 3 **Measure** 50 cm³ of hot water and record the initial temperature of the hot water.

STEP 4 Transfer the hot water to the insulated cup. Stir the water so that the mixture reaches an equilibrium temperature.

STEP 5 Record the final temperature of the mixture.

STEP 6 Calculate the heat or thermal energy involved. Remember, since the density of water is 1 g/cm³, 1 cm³ of water has a mass of 1 g.

STEP 7 Comment on the amount of heat lost by the hot water and the heat gained by the cold water.

STEP 8 Explain why it was important to use an insulated container in this experiment.

STEP 9 Compare the value from procedure A with that you obtained for procedure B.

STEP 10 Compare the value from procedure A with that you obtained for procedure C.

🌐 **GLOBAL CONTEXTS**
Scientific and technical innovation

🧠 **ATL SKILLS**
Communication
Use and interpret a range of discipline-specific terms and symbols.

Activity 6 — Investigating a factor affecting the final temperature of a reaction mixture

Some chemical reactions are exothermic while other reactions are endothermic. There are several factors that affect how much energy is absorbed or released by the reaction mixture.
In this activity, you will design and carry out an experiment to answer the following question:

How does the mass of baking soda used in a reaction with vinegar affect the final temperature of the reaction mixture?

STEP 1 State the chemical equation for this reaction.

STEP 2 List your variables, including those you will fix or control.

STEP 3 **Design** a complete method that will allow you to measure the effect of the mass of baking soda on the final temperature of the reaction. Explain how you will manipulate the variables, and how you will collect data.

STEP 4 Perform a trial test.

STEP 5 Modify your procedure based on your trial and suggest improvements to your procedure.

STEP 6 Collect sufficient data that will enable you to verify your conclusion.

STEP 7 Plot your data on an appropriate graph so that the relationship between mass of baking soda and temperature change is easily shown.

STEP 8 Use scientific reasoning to explain any trends that you observed.

TIP

When designing a scientific investigation, you must determine one variable to change. This is called the **independent variable**. In order to design a valid experiment, you must keep all other variables constant. There should also be one variable that you will be measuring; which is the **dependent variable**.
It is essential that you consider the **control variables** when devising a method. One way to do this is to list all the possible variables that can be changed in an experiment. After determining the independent and dependent variables, all the other variables on the list must be controlled.

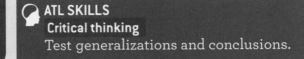
Reflection on Topic 2

Thermochemistry is the area of chemistry that deals with the study of energy changes that accompany chemical reactions and physical changes such as dissolving or changes of state.

- How is the study of thermochemistry relevant to your daily life?

- Consider some of the situations that you come across in your daily life that are related to the transfer of energy in chemical reactions.

TOPIC 3

Sustainable energy resources

The search for sustainable sources of energy remains one of the top global priorities as petroleum reserves are used up. Every human activity requires the consumption of some form of energy—whether from the chemical energy stored in our food, or from energy sources used to generate electricity. Renewable energy is derived from natural sources such as wind, solar, hydro, geothermal, tidal and biofuels. These sources are renewable and sustainable because they will never run out or can be replaced within a human lifetime.

The global population is increasing and with this our demand for energy. It is important that scientists find sources of energy that are clean and efficient.

We know that energy can be converted into different forms, however not all energy is transferred in useful ways. When generating electricity, some of the input energy is transferred to the surroundings as thermal energy. This form of energy is not useful in these circumstances. Energy efficiency is the proportion of the energy supplied that can be usefully used.

Biofuels

Fossil fuels are the major source of energy across the globe. Fossil fuels are a non-renewable resource. This means that their supplies are limited. The main types of fossil fuel that industry and the general population use are coal (used largely in power plants), petroleum (used to fuel automobiles) and natural gas (used to cook food).

CHAPTER LINKS

In Chapter 7 on transformation and Chapter 11 on form, you will find out more about how energy can be transformed from one form to another. For example, nuclear or mechanical energy can be transformed to kinetic energy in a turbine, which generates electricity. In Chapter 7 on transformation, you will also learn more about conservation of energy.

Limited supplies are not the only problem. Fossil fuels produce large amounts of carbon dioxide and other pollutants when combusted. There are many alternative energy resources in development that, it is hoped, will reduce or replace our dependence on fossil fuels.

One alternative is biofuel. Biofuels are made through the process of carbon fixation by plants or microorganisms. Carbon fixation means converting inorganic carbon in atmospheric carbon dioxide to organic compounds (compounds containing C–H bonds). These organic compounds can then be used as fuels to supply energy by combustion.

Biogas is already produced from the breakdown of farm waste or other organic waste. Biodiesel is produced from plant or vegetable oils. New forms of biofuel are being researched. These include bacterial biofuels, algal biofuels and bioethanol. Bioethanol is produced from the plant structural materials, cellulose and lignin, that constitute most of the mass of plants. Bioethanol is made by fermenting the carbohydrates in sugar or starch crops such as maize or sugar cane.

 Activity 7   **What plant source is the best source of oil for biofuels?**

Oil can be extracted from a range of vegetables and crop plants and chemically converted to biodiesel to burn in automobiles. Soybeans are currently a major source of oil for biodiesel. In this activity, you will plan a method to investigate what plant is the greatest source of oil that could be used as a future source of biodiesel.

STEP 1 With a partner, brainstorm a list of plants that could be possible sources of oil for use as biofuels.

STEP 2 One quick way to extract oil from a plant source is to grind the food with a pestle and mortar and then use a centrifuge or a settling technique to separate the oil phase from the water phase.

a) Outline a procedure that makes use of a pestle and mortar and a centrifuge and that allows you to collect enough relevant data to state which plant source would be the best source of oil for the production of biodiesel.

b) List the variables that you will have to consider.

c) **Formulate** a hypothesis that relates to this experiment.

d) Describe what type of data you would have to collect in order to conclude what food source would be the best primary source of a biofuel.

e) How much data would you have to collect to ensure that any conclusions you make are valid?

f) What type of table could you use to display your data so that it is easily understood?

STEP 3 Have your procedure approved by your teacher, make any changes that are needed then carry it out.

STEP 4 Comment on the statement: "Biofuels are the fuel source for the future."

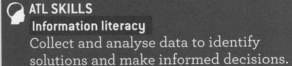

GLOBAL CONTEXTS
Globalization and sustainability

ATL SKILLS
Information literacy
Collect and analyse data to identify solutions and make informed decisions.

Renewable energy sources for power generation

The electricity generation sector is responsible for more than 40 per cent of all carbon dioxide emissions from burning fossil fuels, and about 25 per cent of our total greenhouse gas emissions. It should be a priority for governments worldwide to find alternative fuel sources to decrease the harmful effects caused by burning of fossil fuels.

In 2012, a team from the Carnegie Department of Global Ecology at Stanford University found that wind power has the potential of generating 100 times more power than the global population currently needs (*Nature Climate Change*, 2012). New ideas include the use of airborne and high-altitude turbines, as they can generate more power than ground- and ocean-based turbines due to the higher wind speeds further from the ground.

According to the Global Wind Energy Council, wind turbines are forecast to generate over 416 GW of power in 2015 and reach 600 GW by the end of 2018. The unit of power is the watt (W), and is a rate of working of 1 joule per second. But even by 2035, wind power will only account for 7.5 per cent of global energy production. One major problem with wind power is that it relies on an intermittent source of energy—sometimes it is just not windy. An important area of research is on how electricity generated during windy periods can be stored. Subsequently, electricity could then be sent to the energy grid during non-windy times to maintain a continuous supply of energy.

 Activity 8 Investigating a model wind turbine

You are going to build a model wind turbine and determine the factors affecting the power output of a wind turbine. You will choose one aspect of the wind turbine's design and investigate how this can maximize the power output (the rate at which it does work). You could use your turbine to do non-electrical work such as lifting a weight, or electrical work by lighting a bulb.

The relationship between work done, power output and time is expressed as:

power (in W) = work done (in J) / time (in s)

STEP 1 **Discuss** what modifications are possible in the construction of the wind turbine to increase the power output. Choose one variable to investigate. Explain why you think this modification will increase the power output.

WEB LINKS

Instructions for making a wind turbine can be found by searching an internet search engine for "Wind Turbine Activity science.gc.ca".

For additional projects on wind turbines go to http://learn.kidwind.org/.

STEP 2 Describe how this hypothesis will be tested.

State the independent variable (an aspect of the design you want to change), the dependent variable (what you will measure) and how you will process the data to calculate the power output. List the control variables (all other variables that must stay the same during trials).

STEP 3 List the materials you will need to test your wind turbine. You will have access to a fan.

[SAFETY] Clamp the fan to the bench to prevent toppling. Turn off the fan when making adjustments to the turbine.

STEP 4 Describe a detailed set of procedures for how you will alter the variables and collect data, so that anyone can repeat your experiment.

Have your procedure approved by your teacher, make any changes that are needed then carry it out.

STEP 5 Construct a data table to record the data that you will collect, and include the units for the measurements.

Choose a suitable chart type to present your data to make the findings clear.

STEP 6 After you have collected your data, consider your hypothesis and describe how the independent variable you selected altered the power output of the wind turbine. **Evaluate** the validity of the method, accuracy of the data and any possible improvements or **extensions to the method** for further inquiry.

GLOBAL CONTEXTS
Globalization and sustainability

ATL SKILLS
Critical thinking
Interpret data.

QUICK THINK

After completing Activity 8, explain how what you have found out could help design a more effective wind turbine.

Fusion power

Fusion is the source of energy in stars, like our Sun. As described in the chapter introduction, scientists are investigating how to control a similar fusion reaction here on Earth, to generate electricity. Fusion is theoretically an excellent source of power, and offers the possibility for the world to replace a significant proportion of its energy needs with a safe, clean and virtually limitless resource.

However, it is extremely difficult to create the conditions for fusion to occur. High temperatures and pressures are needed to cause the nuclei of two atoms to overcome the repulsion of the two nuclei due to their electrical charge (both are positively charged). However, if the nuclei are travelling fast enough, they have enough energy to overcome the force trying to keep them apart. When they fuse, they release energy.

Difficulties with fusion power on Earth

Experiments have been carried out at several research reactors to force nuclei together at very high temperature and pressure in a plasma or ionized gas. The plasma is heated to the temperatures needed for fusion, 100 million°C. The plasma is usually controlled in a toroidal (doughnut-shaped) chamber by rapidly compressing it to very high densities using lasers.

The difficulty is in reaching the "break even" point, at which a fusion reactor produces as much energy as is required to keep the reactions going at very high temperatures. Even when this is achieved, it will still be many years before the first commercial fusion reactor is built to reliably provide electricity.

Activity 9   Marshmallow fusion

You are going to use marshmallows as an analogy for nuclear fusion.

STEP 1 Try to fuse two marshmallows by crushing them together.

STEP 2 Observe what you see and **measure** the mass of the marshmallows.

STEP 3 **Measure** the mass of a beaker.

STEP 4 Then try to fuse the marshmallows by heating them in the weighed beaker.

STEP 5 Record your observations.

STEP 6 **Measure** the mass of the marshmallows and beaker after heating.

STEP 7 Find the new mass of the marshmallows.

STEP 8 Discuss what you observed and how this can be used as a model for fusion. What are the strengths and weaknesses of this model?

> ∞ **CHAPTER LINKS**
> In Chapter 8 on models, you will learn more about evaluating scientific models.

🌐 **GLOBAL CONTEXTS**
Scientific and technical innovation

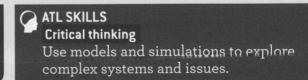
💭 **ATL SKILLS**
Critical thinking
Use models and simulations to explore complex systems and issues.

Future energy supply

Oil prices have risen greatly in recent decades. This has affected how many countries look at their sources of energy. For example, there is the question of whether to invest in renewable power or build new nuclear power stations. There are other factors to consider, such as how to balance costs and benefits.

The EU introduced several targets for member states in 2007. These include the following.

- A cut of 20 per cent in greenhouse gases (eg carbon dioxide, methane) from primary energy sources such as coal and oil by 2020, compared to 1990 levels.

- A minimum target of 10 per cent for the use of biofuels within road transport fuel consumption by 2020.

As governments in the EU plan a long-term sustainable energy strategy, they must take these proposals on board.

QUICK THINK

Find out more about energy proposals that the EU has introduced in the past 10 years. Discuss if these proposals have had the desired effect in the EU.

Africa has a growing economy and will have more industry. With a rising population, rising urbanization rate and greater energy demand from industry, Africa's energy demand is expected to rise steadily.

Develop a flow chart to show how a growing economy, a growing population and more manufacturing will affect the energy demands that the continent has.

⬭ WEB LINKS

On 11 March 2011, a 15-metre tsunami disabled the power supply and cooling of three Fukushima Daichi reactors, causing a major nuclear accident. This disaster refuelled the nuclear power debate worldwide. Learn about the controversy surrounding nuclear power from the following resources and make your own informed decision about this source of power.

Union of Concerned Scientists www.ucsusa.org/nuclear_power/.

Debate on nuclear power: The Economist www.economist.com/debate/overview/201.

Nuclear Power: The Pros and Cons by Ewan McLeish. The Rosen Publishing Group 2007.

Reflection on Topic 3

- What evidence can you find to show that all nations are taking responsibility to cut greenhouse gas emissions?

- Do you think it is viable to harness the energy produced by the Sun? Explain your answer.

Summary

You have learned that energy is stored in food and fuels, is released in chemical reactions including combustion, and can be transferred from renewable sources to generate electricity. Humans spend a lot of time trying to find ways to get stored energy to do useful work before it is transformed into less useful forms and makes its way into the environment as heat.

Perhaps in your lifetime scientists will determine a way to get past the difficulties with fusion technology and we can work out how to harness the power of fusion reactions here on Earth.

TAKE ACTION
Investigate the energy resources that provide power to your school building. What alternative energy sources could be chosen, and which would be more sustainable and suitable for your school?

Evidence

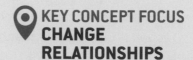

**KEY CONCEPT FOCUS
CHANGE
RELATIONSHIPS**

INQUIRY QUESTIONS

TOPIC 1 Reasoning from evidence

- **What are the features of a testable statement?**
- **What makes a hypothesis useful?**
- **What counts as a fair test of a hypothesis?**

TOPIC 2 The role of evidence in chemistry

- **How could evidence from chemistry experiments be used to control the rate of a reaction to make it safer?**
- **Should governments and science be investing in technologies that could be potentially harmful to the environment or humans?**

TOPIC 3 Big Bang theory—looking for evidence

- **What can we learn about the origin of the Universe from observations of distant stars?**
- **How can we use experiments on Earth to understand the way the Universe is structured?**

SKILLS

ATL

✓ Use brainstorming and visual diagrams to generate new ideas and inquiries.

✓ Make guesses, ask "what if" questions and generate testable hypotheses.

✓ Give and receive meaningful feedback.

✓ Process data and report results.

✓ Consider ideas from multiple perspectives.

✓ Practise observing carefully in order to recognize problems.

✓ Apply skills and knowledge in unfamiliar situations.

Sciences

✓ Formulate testable questions and hypotheses using scientific reasoning.

✓ Explain how to manipulate variables, and how enough data will be collected.

✓ Process data and plot scatter graphs with a line of best fit to identify relationships between variables.

✓ Interpret data and explain results using scientific reasoning.

OTHER RELATED CONCEPTS

**Consequences   Conditions   Environment
Models   Pattern**

GLOSSARY

Fair test an investigation in which only the independent variable has been allowed to affect the dependent variable.

Hypothesis a tentative explanation for an observation or phenomenon that requires experimental confirmation; can take the form of a question or a statement.

Prediction an expected result of an upcoming action or event.

COMMAND TERMS

Comment give a judgment based on a given statement or result of a calculation.

Construct display information in a diagrammatic or logical form.

Introducing evidence

You may be familiar with Charles Darwin's famous voyage aboard the HMS (His Majesty's Ship) *Beagle*. This was an extremely important event in the history of science. On the *Beagle*'s journey to survey the coast of Tierra del Fuego at the southern tip of South America, Darwin's role was to survey the geology, animals and plants of the continent and islands. En route, he visited the Galapagos Islands off the coast of South America. There he first saw the variety of finches that inhabit these islands (Figure 6.1). The observations from this voyage led him to the realization that diversity on Earth arose from a process known as "evolution", which is driven by another process we now know as "natural selection".

Darwin did not realize the significance of his observations straight away, or any time soon afterwards. The voyage took many years and apart from a vague idea that species might not be fixed, Darwin was no closer to a theory of evolution when he returned to England than when he left.

What he did have, however, was a large amount of evidence to support his ideas. This evidence was made up of a vast collection of plants, animals and fossils along with many notebooks of drawings and records of his observations offering possible interpretations of the data collected. It was this evidence that helped to form the theory on the evolution of species that Darwin developed slowly over the 20 years before his publication of *On the Origin of Species*.

Evidence comes in many forms. No matter how we obtain it, evidence assumes a central and essential place in the scientific method, providing support for proposals derived from observation and interpretation of data.

> *It was evident that such facts as these, as well as many others, could be explained on the supposition that species gradually become modified, and the subject haunted me.*
>
> Charles Darwin

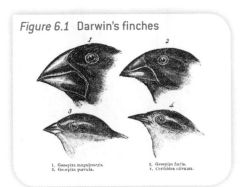

Figure 6.1 Darwin's finches

1. *Geospiza magnirostris.*
2. *Geospiza fortis.*
3. *Geospiza parvula.*
4. *Certhidea olivacea.*

TOPIC 1

Reasoning from evidence

What we know about the natural world is continually changing due to the investigations of scientists. Existing scientific ideas can be updated by adding new knowledge, or can be proved to be wrong.

Not all claims about the natural world are scientific. In order for a statement to be classified as scientific it must be both testable and falsifiable. If a statement is falsifiable, then it must be possible to develop tests that could potentially prove the statement wrong.

Scientists use two kinds of process to investigate the natural world. Each process uses a different method of reasoning.

- In discovery investigations, observations are used to make a generalization (an attempt to identify a general pattern), **hypothesis** or preliminary answer. Repeated observations and patterns are used to form conclusions. This involves inductive reasoning. Reasoning by induction can be thought of as forming generalizations from specific observations.

- In hypothesis-based science, generalizations are used to formulate a hypothesis and to make **predictions**. More specific experiments are then performed to test the hypothesis. This involves deductive reasoning. Reasoning by deduction can be thought of as applying rules to reach a logically certain conclusion.

There are four criteria you can use to evaluate generalizations.

- **Amount of evidence** All of your investigations should obtain sufficient relevant data.

- **Consistency** If your generalization is surprising, then the strength of your evidence is more likely to be closely examined. For example, a claim that plants grow faster when in an environment where music is played requires strong evidence as it is difficult to imagine a mechanism by which this could occur. The generalizations that you make from the evidence should be consistent. This means that you should not be able to draw two different conclusions from the same evidence.

- **Exceptions** You should look to find examples where the generalization does not apply. If such an exception can be found, then the generalization should be refined because it is too general. For example, the generalization that "all matter contracts when cooled" should be refined to "most matter contracts when cooled except for water, which expands when it freezes".

- **Variety** You should consider if your generalization could apply to a variety of scenarios. For example, you might observe that the leaves of maple trees at the edge of a forest are more likely to be eaten by herbivores than the leaves of maple trees inside the forest. You should expand your study to other tree species before forming the generalization that forest-edge trees in general are more likely to be eaten by herbivores.

Testable statements

In this activity, you will investigate the features of testable statements.

The following table compares testable statements with difficult-to-test statements.

Testable statements	Difficult-to-test statements
Divers with high resting heart rates exhaust their scuba tank faster.	Divers who are nervous exhaust their scuba tank faster.
Unlike the vertebrate eye, an octopus eye is focused through movement, like the lens of a camera, rather than by changing shape as the lens in the vertebrate eye does.	The human eye and octopus eye are so complex that they cannot have evolved without the intervention of a divine being.
Handling frogs causes warts.	Handling frogs makes you happy.
Fish gills open and close with greater frequency when water temperature is raised by 10°C.	Fish are affected by temperature.
The class frog spends a greater amount of time on the left side of the tank rather than the right side of the tank when a computer screen plays videos near the left side of the tank.	The class frog likes to watch YouTube.

STEP 1   In a group, compare the different types of statement. Decide what evidence you would require to support each statement and how you would obtain it.

STEP 2   Draw a Venn diagram using the following as a template:

a) In the testable circle, list what the testable statements have in common. In the difficult-to-test circle, list what the difficult-to-test statements have in common. In the intersection you could make a list of things they both have in common.

b) Now, try to use this diagram to develop a set of criteria that could be used to determine if a statement is testable.

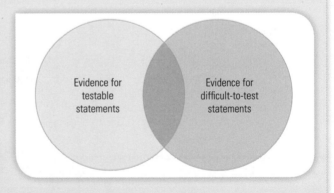

STEP 3   Using the criteria that you have developed, convert the following statements into testable statements:

a) Plants respond to the addition of fertilizer.

b) Brine shrimp need salt to hatch.

c) Dogs are smarter than cats.

d) An earthquake will destroy Toronto in 2019.

e) *Homo sapiens* were better hunters than *Homo neanderthalensis* and that is why Neanderthals became extinct.

f) There are more birds in the sky right now than airplanes.

GLOBAL CONTEXTS
Orientation in space and time

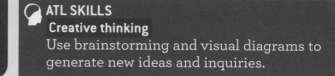
ATL SKILLS
Creative thinking
Use brainstorming and visual diagrams to generate new ideas and inquiries.

The features of a useful hypothesis

Useful hypotheses have the following characteristics.

- Hypotheses are testable by making observations or performing experiments.

- Hypotheses allow predictions to be made. Predictions are commonly expressed as "If …, then …" statements such as "If the independent variable is changed in this way, then the dependent variable will be observed to change in this way."

- Hypotheses suggest a causal link between two things and offer an explanation for observations. Hypotheses can therefore be worded as "If …, then …, because …" For example, "If the height of the individual is increased, then the length of time they can sustain hanging from a bar in a fitness test will decrease, because although muscle strength increases with height it is not in direct proportion to the increase in the mass the taller person has to support." There must be a clear link between the variables. For example, it may be true that children's vocabulary increases with their shoe size, but it is not a plausible hypothesis that knowing more words makes your feet grow; there is no causal link. There is only a causal link between age and size of vocabulary, and between age and shoe size.

- Multiple hypotheses should be considered whenever possible, in order to avoid bias in investigations. Scientists should consider hypotheses other than what they think to be the case. Experiments should be designed to test one hypothesis at a time.

- Hypotheses can be falsified (proven wrong), but they cannot be proven correct. They can only be supported or validated.

Activity 2 — Developing hypotheses

Complete the table to develop a hypothesis for each of the questions listed in the left-hand column. This may require some research. An example has been provided to get you started.

Question	If the (measured, or independent variable) is changed (in this way)	Then the (dependent variable) will change (in this way)	Because (use scientific reasoning to explain the hypothesis)
How does shade affect the growth of grass?	**If the** area is shaded	**then the** dry biomass of grass per unit area that grows over one week will be lower compared with sunny areas	**because** there will be less photosynthesis in the shade and therefore a smaller increase in biomass over one week.
How does drinking coffee affect your heart rate?			
How does distance from the path affect the percentage of leaves that have been predated?			

 **GLOBAL CONTEXTS**
Scientific and technical innovation

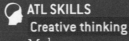 **ATL SKILLS**
Creative thinking
Make guesses, ask "what if" questions and generate testable hypotheses.

Controlling other variables

When biologists claim to know a certain thing about the natural world, how is that claim supported? A hypothesis must be tested by changing a single measured or independent variable and monitoring the outcome, response or dependent variable. All other variables are kept constant or controlled (**fair test**).

The subject matter of biology does vary. In some subfields of biology such as ecology, the response of an entire system is often observed as the measured variable. Other subdisciplines of biology follow a more reductive approach. This means that the hypothesis is tested against fewer variables or perspectives, or only one, where it is possible to do so. However, biological systems are very complex and it is often difficult to control all the variables.

Choose one of the following hypotheses and design a complete and safe method for testing it.

a) If the height of the individual is increased then the length of time he or she can sustain hanging from a bar in a fitness test will decrease. This is because muscle strength increases with height but not in direct proportion to the increase in the mass the taller person has to support.

b) If an area of grass is shaded then the dry biomass of grass per unit area that grows over one week will be higher compared with grass grown in sunny areas. This is because in the shade there will be less transpiration (loss of water from a plant in the form of water vapour, through pores in the leaves) leading to more carbon dioxide uptake, leading to more photosynthesis and therefore a greater increase in biomass over one week.

Ensure you describe how only the independent variable will be changed, how the dependent variable will be measured and how other variables that could have an influence on the value of the dependent variable will be kept constant. When you have completed your plan, ask a partner to evaluate your method.

 **GLOBAL CONTEXTS**
Scientific and technical innovation

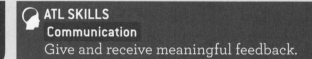 **ATL SKILLS**
Communication
Give and receive meaningful feedback.

Reflection on Topic 1

- Is every question scientific?

- What qualities should a question have if we are to test it using scientific methods?

- What makes a hypothesis different from a prediction?

- How can we make sure our hypothesis is being tested?

WEB LINKS
Evidence is key in science but how do we know it is not biased? What happens when different groups have an interest in a scientific discovery? Does this alter the type of evidence collected? Visit the University of California at Berkeley website at www.berkeley.edu and search for "who pays for science?".

TOPIC 2

The role of evidence in chemistry

Scientific discovery is based on recording evidence and making conclusions based on this evidence. It is important not to take everything that is reported in the media and online as scientifically correct truth; much may be unsupported by evidence. Scientific truth is based on evidence that we can observe directly or that we collect via technology. Because a lot of chemistry happens at the atomic or subatomic level, much of the evidence in chemistry is collected by observing 'macroscopic' properties of matter such as mass or volume during experiments.

Rates of reaction

Dynamite explodes. Iron rusts. These are two examples of chemical reactions, but they are very different. The explosion of dynamite represents a very, very fast reaction; it is over in a few seconds. On the other hand, the rusting of iron on your bicycle frame or your parents' car is very slow and takes months or years to show a noticeable difference. The main difference between these two chemical reactions is the rate (how fast or slow) at which they occur. A study of the rates of chemical reactions is called chemical kinetics. Reaction rates can be measured experimentally when chemists collect evidence during reactions.

Chemists explain rates of chemical reactions by means of the collision theory. The collision theory states that:

- in order for a chemical reaction to occur, a collision between the reactant atoms or molecules must occur

- in order for a collision to be successful, the reacting particles must collide with a minimum amount of energy

- in order for a successful collision to occur, the colliding molecules must hit each other the right way (have the proper orientation of their shapes).

QUICK THINK

There are many examples of controlling reaction rates in everyday life.

a) Which of the following makes it easier to start a campfire—a large dry log or small pieces of dry kindling?

b) Why does fruit spoil more quickly at room temperature rather than when it is in the fridge?

Search for "fruit vegetable decomposition time lapse" on www.youtube.com.

WEB LINKS

To find some informative and entertaining videos about collision theory, search for "collision theory explained" on www.youtube.com.

 Activity 4   **Factors affecting reaction rates**

A rate of reaction can be determined experimentally by monitoring the change in the amount of reactant used up, or the change in the amount of product formed over a period of time.

Evidence collected during this experiment will allow you to comment on how changing the variables in a reaction can alter its reaction rate. You will be changing one variable at a time for the reaction between hydrochloric acid and calcium carbonate.

You may not be familiar with this reaction. It is one of the reactions that takes place when acid rain damages marble statues and buildings constructed from limestone or marble. You will investigate this reaction in more detail in chapter 10.

STEP 1   Gather together the following materials:

- eye protection
- 250 cm³ Erlenmeyer or conical flask
- plug of cotton wool for the top of the conical flask
- balance
- 50 cm³ and 100 cm³ graduated cylinders

- small lumps of calcium carbonate
- 2M hydrochloric acid
- a water bath with the temperature set at 40°C.

STEP 2 Write the complete chemical equation for this reaction.

STEP 3 **Construct** a data table to record the data that you and/or the class will collect.

STEP 4 **Procedure A**

[SAFETY] Hydrochloric acid is an irritant: avoid contact with eyes, mouth and skin. Wear eye protection at all times.

a) Place a 250 cm³ Erlenmeyer or conical flask on a balance and add about 10 g of small lumps of calcium carbonate.

b) Carefully measure out 50 cm³ of 2M hydrochloric acid using a graduated cylinder.

c) Place this graduated cylinder alongside the Erlenmeyer flask on the balance. Measure and write down the total mass.

d) Add the acid to the flask while it is still on the balance. Plug the top with the cotton wool. Measure and record the total mass at 30-second intervals until the mass stops changing.

STEP 5 **Procedure B**

Repeat Procedure A, but use 100 cm³ of 2M acid rather than 50 cm³. The variable being investigated is how a greater volume of a reactant will affect the rate of reaction.

STEP 6 **Procedure C**

Repeat Procedure A, but use 25 cm³ of 2M acid and 25 cm³ of water instead of 50 cm³ of acid. The variable being investigated is how a reduced concentration of acid will affect the rate of reaction.

STEP 7 **Procedure D**

Repeat Procedure A, but use 5 g of calcium carbonate instead of 10 g. The variable being investigated is how reducing the mass of a reactant will affect the rate of reaction.

STEP 8 **Procedure E**

Repeat Procedure A, but put the flask to warm in the water bath set at 40°C before adding the calcium carbonate. The factor being investigated is how increasing the temperature of a reactant will affect the rate of reaction.

TIP

When drawing your graph:
- ensure that the graph is fully labelled
- choose an appropriate range for each axis
- put the independent variable (the variable you are changing) on the x-axis (there is an exception here if you are measuring time; this always goes on the x-axis)
- a best-fit trend line is a smooth curve; draw the curve so that about half the points are above the curve and about half the points are below it.

Procedure F

Repeat Procedure A, but use one large lump of calcium carbonate, rather than a number of small lumps. The factor being investigated is how reducing the surface area of a reactant will affect the rate of reaction.

STEP 10 Calculate the loss in mass of the reaction mixture for each of the procedures.

Plot the loss in mass versus time for each of the procedures A–F on the same graph. Use different symbols to plot the points for each procedure. Add a best-fit trend line for each procedure. **Comment** on what the lines on the graph show.

STEP 11 a) Explain why procedure A was important.

b) Discuss the factors that affected the rates of this chemical reaction.

WEB LINKS
A catalyst can also affect the rate of a chemical reaction. Search for "decomposing hydrogen peroxide using a catalyst" on www.youtube.com.

GLOBAL CONTEXTS
Scientific and technical innovation

ATL SKILLS
Information literacy
Process data and report results.

A reaction rate is a measure of the speed of transformation of reactants to products in a chemical reaction. The factors that affect the rate of reaction are the state, concentration or surface area of the reactants, the temperature at which the reaction is taking place or the addition of a catalyst. These factors determine conditions that either increase or decrease the number and frequency of collisions between the reactant molecules.

If the solution remains at a constant temperature, the frequency of collisions depends only on the number of reacting molecules present in a given volume of reacting material. An increase in the concentration of reacting substances increases the frequency of collisions.

Activity 5   Iodine clock reaction

Before you begin this activity, review the conclusions you came to in the previous activity.

In this complex reaction, which happens between several different reactants, the concentration of one reactant (potassium iodate) will be changed one drop at a time to see the effect it has on the reaction. You will use a different form of evidence—a colour change—to show when the reaction is complete.

STEP 1 Gather together the following materials:

- eye protection
- two 50 cm³ beakers
- 25 cm³ graduated cylinder
- droppers
- distilled water
- potassium iodate solution
- sodium metabisulfite solution
- concentrated sulfuric acid
- starch solution
- stopclock/timer.

STEP 2 **[SAFETY]** Wear eye protection at all times. Concentrated sulfuric acid is very corrosive. Use only one drop at a time. Avoid contact with eyes, mouth and skin.

a) Obtain two 50 cm³ beakers. Make sure that they are clean.

b) To one beaker add 10 cm³ of distilled water, followed by a number of drops of potassium iodate. (Your teacher will tell you how many drops.)

c) To a second 50 cm³ beaker add 25 cm³ of distilled water, 3 drops of sodium metabisulfite, 1 drop of concentrated sulfuric acid and 10 drops of soluble starch.

d) Quickly pour beaker 2 into beaker 1, starting the timer as soon as the two solutions meet.

e) Pour the liquid back and forth between the two beakers 2 or 3 times to ensure mixing.

f) At the moment a colour change occurs, stop the timer and write down the time to the nearest tenth of a second.

g) Repeat the experiment for different numbers of drops of potassium iodate, as instructed by your teacher.

STEP 3 Complete the following tasks:

a) Describe the relationship between concentration of potassium iodate and time taken for the colour change.

b) Suggest some other ways that you could change the reaction rate in this experiment.

c) Plot the number of drops versus time on a graph, and draw a line of best fit.

d) Suggest one possible source of error when recording the time for the colour to change.

e) Suggest how the method could be improved to reduce the inaccuracies in the method.

GLOBAL CONTEXTS
Scientific and technical innovation

ATL SKILLS
Information literacy
Process data and report results.

Chemical explosions

Chemical explosions are examples of reactions that occur very quickly as a result of a rapid production of gases and the release of energy. One of the pioneers in the field of chemical explosives was Alfred Nobel, the inventor of dynamite. Dynamite is widely used in the construction industry to blast through rock so roads can be laid, or in mining to expose sources of minerals. Dynamite was also widely used in World War I and resulted in tens of thousands of deaths. This would have been very distressing for Alfred Nobel had he still been alive, because he was a pacifist.

Scientists have to continuously evaluate the worth of their research. The evidence that they collect can lead them to new discoveries, further experimentation and the development of new products or processes. The science of explosives is still explored today, even though their use can cause harm to users or to the environment. Experiments on different materials give evidence for how the same explosive power can be achieved with materials that are safer to handle and store. Researchers also look for explosives that do not release toxic gases into the atmosphere or leave toxic residues in the fallout. Research into rates of reaction could help control industrial chemical processes where a sudden release of energy would be dangerous.

 **WEB LINKS**
Explore the power, magic and chemistry of explosives. Search "PBS Nova Kaboom!" to find out more about the history of explosives, from the secret experiments of early alchemists to the lethal legacy of Los Alamos.

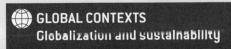 **Activity 6**   **Exploring the ethics of explosives**

Answer the following questions and share your responses with a partner or the larger group:

a) Comment on Oppenheimer's statement opposite.
b) Based on evidence that you observed in the video, discuss whether funding should be provided to scientists to investigate new forms of explosives.
c) What is the responsibility of scientists who know the research they are involved with will result in harm to the environment or humans? Should they be held responsible for these discoveries? Or should they be congratulated for making new discoveries that could change the world?
d) Discuss and evaluate the implications of the use of improved chemical knowledge in the development of potentially harmful chemicals.

When it [the atomic bomb] went off in that New Mexico dawn, that first atomic bomb, we thought of Alfred Nobel, and his hope, his vain hope, that dynamite would put an end to all wars.

Robert J Oppenheimer

🌐 **GLOBAL CONTEXTS**
Globalization and sustainability

💭 **ATL SKILLS**
Critical thinking
Consider ideas from multiple perspectives.

Reflection on Topic 2

Chemical kinetics is not the only area of chemistry where evidence plays an important role. Through observation, evidence can be gathered to help explain how changing the conditions of a reaction can affect reaction rates. Evidence is key for scientists to validate their hypotheses.

When thinking about evidence and experimentation, consider the following questions:

- How much evidence is enough to consider an experiment a success?

- What are the best types of evidence that a chemist can collect?

- Does all evidence a chemist collects have to be collected experimentally or can evidence be obtained through computer models and simulations?

TOPIC 3

Big Bang theory—looking for evidence

In order to support scientific theories and laws, there must be sufficient and unbiased evidence.

Evidence can be used to gradually develop a theory as more evidence is collected. For example, in order to develop a theory about the structure of the atom, scientists gathered evidence from many experiments. JJ Thompson discovered negatively charged particles, later called electrons. A few years later, Ernest Rutherford and his students found evidence for a small, positively charged nucleus with his "gold foil experiment" in 1909 (Figure 6.2).

> *It was almost as incredible as if you fired a 15-inch shell at a piece of tissue paper and it came back and hit you.*
>
> Ernest Rutherford

WEB LINKS

Animations of Rutherford's famous gold foil experiment can be found at www.learnerstv.com by searching for Rutherford animation and micro. magnet.fsu.edu by searching for Rutherford experiment.

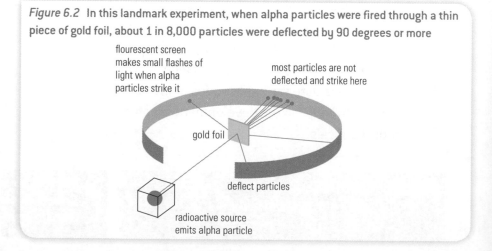

Figure 6.2 In this landmark experiment, when alpha particles were fired through a thin piece of gold foil, about 1 in 8,000 particles were deflected by 90 degrees or more

flourescent screen makes small flashes of light when alpha particles strike it

most particles are not deflected and strike here

gold foil

deflect particles

radioactive source emits alpha particle

Using the scientific method, we can make observations and test our own hypotheses against these observations in a search for theories to help explain each phenomenon. Scientists must employ critical thinking skills and perseverance to analyse these observations and construct supporting evidence for a convincing scientific theory.

Origin of the Universe

Until around 1920, astronomers did not know the size of the Milky Way galaxy, and debated whether other galaxies existed. The best estimate of the size of the Milky Way was 300,000 light years across. A light year is the distance that light travels in one year: it is 9.4×10^{15} metres (or 9,400,000,000,000,000 metres). In 1924, American astrophysicist Edwin Hubble found that the distance to the Andromeda nebula, a hazy spiral visible in the night sky with a good telescope, was over a million light years. He also found that there were individual stars in the nebula. He proposed that there were distant galaxies outside of our own Milky Way, each containing billions of stars.

We now know that the Universe contains millions and millions of galaxies other than our own. We also have evidence for the beginning and evolution of the Universe from a single point—a very small, very dense and very hot point called a singularity. All the matter that makes up our Universe was in this one place at the beginning of the Universe. An explosive event caused matter from the singularity to spread out, forming the Universe we inhabit today. This is the Big Bang theory, which is the currently accepted model for the origin of the Universe. It is estimated that the Big Bang took place over 13.8 billion years ago, and the Universe is still expanding today, with galaxies moving away from each other at up to a third of the speed of light.

One of the key pieces of evidence for the Big Bang theory is the expansion of the Universe. This is supported by data from a phenomenon called redshift, also discovered by Edwin Hubble. You will investigate the Doppler effect in sound as an analogy for the Doppler effect or redshift in light waves.

Observing distant stars

The visible light from stars is part of a continuous spectrum, from infrared to ultraviolet. Some objects in the Universe give out radio waves or gamma rays. These types of electromagnetic radiation are all part of the electromagnetic spectrum, which is continuous, so all values of wavelength are possible (Figure 6.3).

 **CHAPTER LINKS**
See Chapter 8 on models for a comparison of Thompson's and Rutherford's models of the atom.

WEB LINKS
Find out more about Edwin Hubble by exploring hubblesite.org.

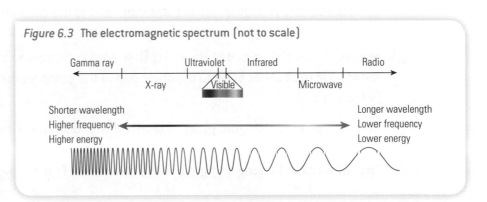
Figure 6.3 **The electromagnetic spectrum (not to scale)**

A wavelength is the distance in metres between repeats in a wave. The longer the wavelength, the more spread out the wave. Frequency is the number of waves that pass by a certain point per second. The longer the wavelength, the lower the frequency. The shorter the wavelength (waves very close together), the higher the frequency (the waves are passing a point more often).

Activity 7   Visualizing the Doppler effect for sound

The purpose of this activity is to investigate the effect the movement of a source of sound has on what you hear as it moves towards and away from you. Sound travels as a wave, but a different type from light waves.

Sounds with longer wavelengths (lower frequencies) are interpreted as lower pitch by our brains. Shorter wavelengths (higher frequencies) would mean the waves are closer together, and this is heard as a high pitch.

> **⌘ CHAPTER LINKS**
> See Chapter 8 on models for a comparison of how light waves and sound waves are represented.

STEP 1 Locate the following materials:

- electronic sound source with a pure tone (examples: small alarm clock with continuous tone, buzzer connected to a cell)
- strong string (2.5 m)
- large rubber bung
- sticky tape.

STEP 2 Formulate a hypothesis for what you think you will hear when someone else swings the sound source around his or her head, so that it moves towards and then away from you. Explain why you think this will happen.

Write your hypothesis using the following questions as your framework:

What do you expect to hear, and why, when a source of sound:

a) moves towards you (imagine the sound of a racing car driving towards you)
b) moves away from you (imagine the sound of a racing car driving away from you)
c) is stationary (imagine you are standing next to the car with the motor running)?

STEP 3 Secure the sound source to the string (use knots and tape to ensure it does not detach from the string). Tie the rubber bung to the other end of the string so that it cannot slip out of the hand holding it.

Select one student in your group to twirl the sound source in a circle above his or her head. In order to gain enough speed, the student should let out about 1.5 m of string as it twirls. The twirler should hang on tightly to the other end of the string.

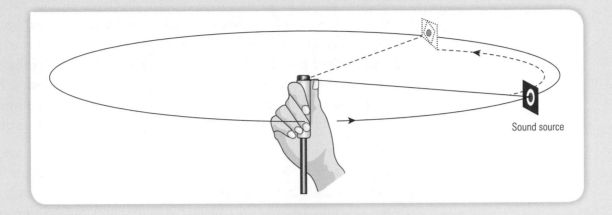

Sound source

[SAFETY] The rest of the group must stand at least 3 metres away from the twirler. If this is not possible in the lab, move to a larger space or outside.

You should observe and record what you hear as the sound source approaches, passes and goes away from you. Each member of the group should have a chance to twirl the device, and also to describe what they hear as other people swing the device.

STEP 4 Evaluate your hypothesis—return to the questions in step 2, answering these again with your new evidence.

STEP 5 **Construct** a data table to record your observations.

STEP 6 The phenomenon you just experienced is called the Doppler effect.
Answer the following questions alone or with your partner:

a) Describe and explain how the sound waves behave when the source is stationary.
b) Describe how the sound changes as the sound source moves towards and then away from you.
c) Use the diagram below to explain how the sound waves from the moving object seem to change pitch. The diagram shows a picture from above of sound waves spreading out from a source. The distance between the wave fronts is the wavelength.

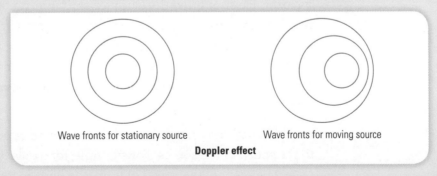

Wave fronts for stationary source   Wave fronts for moving source
Doppler effect

GLOBAL CONTEXTS
Orientation in space and time

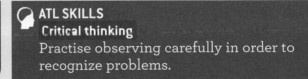
ATL SKILLS
Critical thinking
Practise observing carefully in order to recognize problems.

When a source of waves moves towards an observer, the wavelengths of the waves are closer together, and are shorter. For sound waves, this results in a higher pitch tone. In contrast, as waves move away from an observer, the wavelengths are further away from each other, and are longer. For sound waves, this results in a lower pitch tone. For light waves from a star or galaxy, the analogous effect (as sources move towards and away from the viewer) is a change in the colour of light.

When astrophysicists analyse light emitted by a star, there are gaps or dark lines in the spectrum of light. Not all stars shine in the same way, but there is a characteristic pattern that can usually be recognized. The lines are caused by neutral and ionised elements in the outer, cooler gas layers of the star, which absorb some of the light emitted by the star (Figure 6.4). Different elements absorb light of different wavelengths. When compared to the spectrum of the Sun, the wavelengths of these gaps always appear to be slightly longer or shorter than expected. By 1925, data for 45 galaxies showed that the wavelengths were nearly always longer than expected, towards the red end of the spectrum. Longer wavelengths lead to the appearance of red-shifted light. Shorter wavelengths lead to the appearance of blue-shifted light.

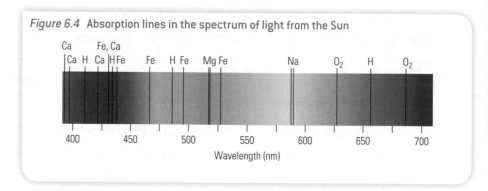

Figure 6.4 Absorption lines in the spectrum of light from the Sun

Edwin Hubble was one of the astronomers studying the light from these galaxies. He supported the Doppler shift interpretation of the observed redshift—that the galaxies were moving towards or away from the Earth, with most of them moving away. As the Sun is not moving towards or away from the Earth, it acts as a reference point and this can be used to help measure how fast stars or galaxies are moving towards or away from the Earth.

Hubble knew that the amount of the observed redshift is proportional to the speed of the source. For example, for a galaxy moving away from us at 3 per cent of the speed of light, the lines in its spectrum will be red-shifted by 3 per cent.

Activity 8   Hubble's evidence for redshift

While studying the stars of distant galaxies, Hubble used another technique to measure the distances to the galaxies. He now had two pieces of evidence for each galaxy. Here is the data he published for 24 galaxies in 1929.

STEP 1   Open a spreadsheet program and enter the data from the table below.

STEP 2   Use your spreadsheet program to create a scatter graph.
Add a *y*-axis title and an *x*-axis title. Include the units.
Add a trendline.

STEP 3   Outline the trend in Hubble's observations shown by the scatter graph.

a) What is the relationship between the distance of the galaxy and its speed relative to the Earth?

b) How does this support the hypothesis that the Universe is expanding?

Galaxy	Distance (in units of 10^6 parsecs)	Measured velocity (km/sec)
small Magellanic cloud	0.032	+170
large Magellanic cloud	0.034	+290
NGC6822	0.214	−130
598	0.263	−70
221	0.275	−185
224	0.275	−220
5457	0.45	+200
4736	0.5	+290
5194	0.5	+270
4449	0.63	+200
4214	0.8	+300
3031	0.9	−30
3627	0.9	+650
4826	0.9	+150
5236	0.9	+500
1068	1.0	+920

5055	1.1	+450
7331	1.1	+500
4258	1.4	+500
4151	1.7	+960
4382	2.0	+500
4472	2.0	+850
4486	2.0	+800
4649	2.0	+1090

A parsec is a unit of distance equal to 3.26 light years. You do not need to convert the distances to light years for this activity.

GLOBAL CONTEXTS
Orientation in space and time

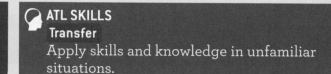

ATL SKILLS
Transfer
Apply skills and knowledge in unfamiliar situations.

Using evidence to develop theories

Hubble had found evidence that suggested cosmology needed a rethink. Scientists need a large amount of data to be convinced, so Hubble was rather cautious when he published his results, writing "New data to be expected in the near future may modify the significance of the present investigation, or, if confirmatory, will lead to a solution having many times the weight. For this reason it is thought premature to discuss in detail the obvious consequences of the present results ...".

After his announcement, Hubble and other astronomers continued to measure the distances and redshifts of thousands of galaxies, to confirm beyond doubt the existence of the redshift to distance relationship.

The galaxies that are furthest away from us are moving the fastest. Since the speed of light is 3×10^8 m/s, light from these very distant galaxies took much longer to reach the Earth. Thus, as you look further out into space, you are in fact looking at light created a long time ago.

TIP

The cosmological redshift is not a true Doppler effect. The light from nearby galaxies is redshifted as a result of the Doppler redshift due to relative velocity of the galaxy and observer, but most distant galaxies are redshifted as a result of space itself stretching, as predicted by Einstein.

Evidence about the Big Bang from the Large Hadron Collider

The Large Hadron Collider (LHC) at CERN allows scientists to reproduce the conditions that existed within a billionth of a second of the Big Bang. These unique conditions allow scientists to confirm the existence of particles predicted nearly 50 years ago. One such particle is the Higgs boson. It is thought to give mass to matter and through this it shaped the Universe 12.7 billion years ago.

The LHC is part of the biggest physics experiment in history. It involves 111 nations working together to design, build and test equipment. Located in Geneva, Switzerland, the LHC is a 27 km ring used to accelerate two high-energy particle beams close to the speed of light, to collide them at very high energies. The collision creates subatomic particles that are not normally observed because they are very unstable and decay into other particles almost immediately.

Reflection on Topic 3

- Have your preconceptions of stars or the Universe been affected by your understanding of the properties of light?

- Is it important that CERN continues to keep asking questions of the Universe?

Summary

You should now have a sense of how evidence, whether intentionally collected, accidentally found or deduced by cutting out other factors, is one of the most vital components of the scientific method. The search for it has occupied the greatest scientific minds for centuries as it provides the foundation on which we can build everything from the simplest hypothesis to the fundamental laws that govern the Universe.

Reference

Hubble, E. 1929. "A Relation between Distance and Radial Velocity among Extra-Galactic Nebulae". *Proceedings of the National Academy of Sciences of the United States of America*. Vol 15, Issue 3. Pp 168–173.

Transformation

📍 **KEY CONCEPT FOCUS
CHANGE
RELATIONSHIPS**

**INQUIRY
QUESTIONS**

TOPIC 1 Energy in food chains

- **How can the maximum biomass of plant matter be generated from photosynthesis in a fixed time period?**
- **How can the maximum biomass of animal matter be generated in a fixed period of time?**
- **How can we modify our diet to reduce our ecological footprint?**

TOPIC 2 Transformation of energy and matter

- **How does energy change from one form to another?**
- **How can we use the transformation of matter as represented by the decay of radioactive isotopes to our advantage?**

SKILLS

ATL

✓ Gather and organize relevant information to formulate an argument.

✓ Collect and analyse data to identify solutions and make informed decisions.

✓ Process data and report results.

✓ Use models and simulations to explore complex systems and issues.

Sciences

✓ Design a method and select appropriate materials and equipment.

✓ Explain how to manipulate variables, and how enough data will be collected.

✓ Organize and present data in tables.

✓ Decide how the raw data should be transformed and presented into a form suitable for visual representation.

**OTHER
RELATED
CONCEPTS**

Energy   Form   Models

GLOSSARY

Accuracy how closely a measurement comes to the true value.

Control variables the variables that remain constant and unchanged in an experiment, to investigate the effect of changing the independent variable.

Dependent variable the variable in which values are measured in the experiment.

Independent variable the variable that is selected and manipulated by the investigator in an experiment.

COMMAND TERMS

Determine obtain the only possible answer.

Present offer for display, observation, examination or consideration.

Introducing transformation

From this quote you can appreciate the tension that characterized the difficult relationship between physicist Nikola Tesla and the inventor Thomas Edison. In the early days of their relationship Edison hired Tesla to help him with some of his ideas, particularly on increasing the efficiency of his direct current (DC) power generation plants. Their relationship quickly soured, partly over money but also because Tesla was developing alternating current (AC) power generation and Edison saw this as competition (Figure 7.1). Their feud was dubbed "The War of Currents".

As you may remember, direct current (DC) flows in one direction from one terminal of the power supply or battery through the loads in the circuit to the other terminal (Figure 7.2). Alternating current (AC) reverses direction periodically—for mains electricity this is usually 50 to 60 times every second.

Transformation means a change in form. In physics this can be defined as a change from one well-defined state to another well-defined state or an alteration in form or condition. This idea of transforming from one form to another brings us to the reason behind the war between Edison and Tesla and what helped bring about the eventual victor.

Both forms of electrical current are capable of lighting our lives and toasting our bread and charging our tablets—the energy carried by moving electrons has the ability to do work in the circuit regardless of the direction of electron flow. Both AC and DC power are also generated with equal ease and efficiency.

However, it is very difficult to transform direct current from one voltage to another, which is a disadvantage when transmitting electrical power from power stations to where it is needed in homes and workplaces. Changing voltage is necessary for electrical transmission and distribution networks because of the energy loss when electrical energy is transmitted through power lines. Whenever a current flows through a conductor, there is a heating effect due to the conductor's electrical resistance. This means energy is lost to the environment. The longer the wire, the more energy is wasted as heat but, more importantly, the power loss is proportional to the square of the current. A smaller current has a much smaller power loss from heat so it is an advantage to transmit power at a low current. But to reduce the current, we must increase the voltage in order to keep the power the same.

This is why electricity generated from the power station is transformed to a very high voltage (110 kV and above) for transmission. Electrical power transmitted at low voltages results in very high energy losses.

If Edison had a needle to find in a haystack, he would proceed at once with the diligence of a bee to examine straw after straw until he found the object of his search ... I was a sorry witness of such doings, knowing that a little theory and calculation would have saved him ninety percent of his labor.

Nikola Tesla

Figure 7.1 Nikola Tesla

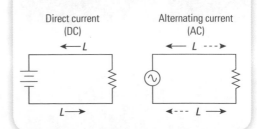

Figure 7.2 Electrons are negatively charged and therefore flow around the circuit from the negative terminal of the cell to the positive terminal. However, the direction of current flow is always shown as from the positive terminal to the negative terminal. This is just a historical convention.

Figure 7.3 A transformer on a post supporting overhead power lines "steps down" the voltage used in power distribution lines to the 110/240 V mains supply used by homes

QUICK THINK

To get a sense of how bitter the rivalry between Edison and Tesla was, do some research about Topsy the Elephant.

It would appear Edison purposely used high voltage alternating current to electrocute the animal to try to show the world that Tesla's method was dangerous.

Figure 7.4 Duckweed

Alternating current can be easily transformed to the extremely high voltages (and subsequent minimal power loss) for transmission and then transformed into voltages that are safe to use in domestic and industrial settings (Figure 7.3). Household supply voltages are between 110 V and 240 V depending on the country. But transformers only work with alternating current. DC voltages cannot be transformed.

Transformation in science is extremely varied, from changes in energy form to changes in physical state. In biology, transformation can involve differentiation of a cell or a molecular level, or alteration of molecules and/or genetic material. The mechanisms behind these transformations are equally varied. It is clear that it is essential to understand this concept if we want to better understand everything from our Universe to the subatomic particles that provide its building blocks.

TOPIC 1

Energy in food chains

Maximizing the formation of plant biomass

The chemical processes that occur within the cells of organisms require energy. Individual organisms exist as part of biological communities. For most biological communities, the initial source of energy is light captured by plants undergoing photosynthesis. A portion of this energy is used by the plant in cellular respiration. Energy stored in plant tissues can contribute to the formation of plant biomass or it can be passed to other organisms through food chains. Energy is passed from consumer to consumer in a food chain, but with every transformation, energy is lost from the community as heat.

Duckweed is the common name given to a small flowering plant that grows on the surface of water in many ecosystems around the globe, even tolerating polluted water. When conditions are ideal, the rate at which a duckweed population can produce biomass is very high compared with other plants.

Fresh or dried duckweed is being investigated as a supplementary animal feed. It has been used in integrated farming in poultry, pork and fish production (Figure 7.4). Integrated farming systems seek to produce a maximum amount of animal feed with minimum resources. This would mean, for example, not using fertilizers.

Activity 1   Investigating optimal conditions for the formation of plant biomass

You are going to design a method for producing the maximum gain in biomass of duckweed over a three-week period. Duckweed is very common in nature but can also be obtained from aquarium suppliers.

STEP 1   Plan how to measure biomass (**dependent variable**) at the start and finish of the experiment.

STEP 2   Since several variables are involved in forming ideal conditions, consider how you will ensure the method is valid. Decide which variable you will alter (**independent variable**) and decide what the **control variables** are. Nutrients, light and warmth are three of the most important factors affecting plant growth but other factors exist as well.

STEP 3   Develop a plan for collecting sufficient relevant data over a three-week period.

STEP 4   Decide how to present your data, which may involve records of several variables over a three-week period. Construct a data table to record the data that you will collect, and include the units for the measurements. Decide how you will display the results in a chart or graph.

STEP 5   Write down the procedure your group agrees on, and have it approved by your teacher.

[**SAFETY**] If infrared or other heat lamps are used, it is important to avoid contact with water. Do not set them up or switch them on or off with wet hands.

Wash your hands after handling duckweed and keep skin contact to a minimum.

GLOBAL CONTEXTS
Globalization and sustainability

ATL SKILLS
Critical thinking
Gather and organize relevant information to formulate an argument.

Feed conversion ratios

Plants capture energy in biomass and this energy is passed along food chains. Less and less energy is available to consumers along the food chain due to losses to the environment as heat and undigested material.

An important variable with respect to the sustainable production of animals for food is the concept of feed conversion ratio. This is defined as the mass of food required to raise animal body mass by one kilogram. The ratio would never be 1:1 because no energy transformation is ever 100 per cent efficient. One proposal for a more sustainable source of animal protein is for humans to use insects as food.

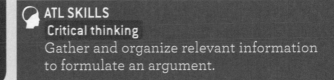

CHAPTER LINKS
See Topic 3 in Chapter 5 on energy for more on the law of conservation of energy and how the efficiency of energy transfers is linked to sustainable energy resources

Activity 2   Comparing mealworms to traditional sources of protein

The graph opposite shows the area of land required to produce one kilogram of edible protein for four traditional foods consumed by humans, and for mealworms (green bar). The red bars indicate the maximum estimates and the blue bars represent the minimum estimates.

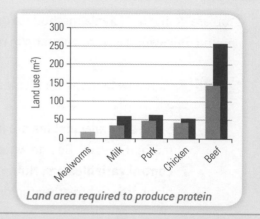

Land area required to produce protein

a) Estimate the area of land required to produce one kilogram of edible protein for mealworms.

b) Evaluate the conclusion that beef is the least sustainable choice of animal protein.

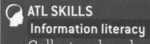 **GLOBAL CONTEXTS**
Globalization and sustainability

ATL SKILLS
Information literacy
Collect and analyse data to identify solutions and make informed decisions.

 Activity 3   Feed conversion ratios in mealworms

You are going to design and carry out an investigation into one of the factors that lead to the maximum gain in biomass of mealworms over a three-week period. For example, you might investigate the effect of different types of food source being fed to the mealworms.

Mealworms can be obtained from a pet supply store. They can be kept in a disposable plastic container along with a food source. The container needs holes for gas exchange.

[SAFETY] Wash your hands after handling the mealworms.

STEP 1 Decide how you will measure biomass at the start and finish of the experiment.

STEP 2 Several variables are involved in deciding ideal conditions. Consider how only the independent variable will be changed and how you will control the other variables.

STEP 3 Develop a plan for how you will collect sufficient relevant data over a three-week period.

STEP 4 Construct a data table to record the data that you will collect, and decide how you will process the results and **present** the data.

STEP 5 Write down the procedure your group agrees on, and have it approved by your teacher.

TIP

When designing your experiment, consider the ethical and environmental implications. Insects gathered from the local environment could be substituted for mealworms as they could be released when the experiment is finished. Ensure that during your experiment the animals are not exposed to conditions that are outside of their limits of tolerance. Because the process of determining dry biomass would kill the insects, you will need to measure wet biomass.

When measuring the mass of the insects, take care not to include any of the food material. Carefully use a small paintbrush to remove the material.

⊕ GLOBAL CONTEXTS
Globalization and sustainability

◉ ATL SKILLS
Information literacy
Process data and report results.

  **Activity 4**   Interpreting your ecological footprint

The area of land required to support the diet, travel, home life and consumption patterns of one person is known as an ecological footprint. It is usually measured in hectares. One hectare is 10,000 m². The average individual living in a developing nation has a much lower ecological footprint than an individual living in a developed nation.

STEP 1 Carry out an ecological footprint assessment using an online tool that allows you to focus on how your diet contributes to your footprint. **Determine** your footprint.

 **∞ WEB LINKS**
Search for "ecological footprint calculators".

STEP 2 Discuss the consequences of altering your lifestyle to reduce your ecological footprint.

STEP 3 Interpret your result to form a conclusion as to how you could modify your diet to reduce your footprint.

⊕ GLOBAL CONTEXTS
Globalization and sustainability

◉ ATL SKILLS
Information literacy
Collect and analyse data to identify solutions and make informed decisions.

Reflection on Topic 1

You have had a chance to think about transformation in food chains and how this knowledge can be used to think about the way your own diet and personal food chain works.

- Will you consider energy transformation when you are selecting your diet?

- Are there alternative sources of food with less impact on the environment that we could consider adding to our diets?

Transformation of energy and matter

Transformation of energy

In our bodies, chemical energy stored in food is transformed by our digestive system into heat to keep us warm and into other stores of chemical energy needed for essential processes such as respiration and growth. The elastic potential energy stored in an elastic band can be transformed into kinetic energy when the band is released. Nuclear power stations transform the nuclear energy stored in uranium into thermal energy, which is then transformed into kinetic energy as the steam drives a turbine.

The law of conservation of energy

The law of conservation of energy states that in a closed system energy cannot be created or destroyed but can only be stored or transferred from one form to another.

This law holds true in all situations, even when it appears that an object has lost energy. For example, a ball moving across a horizontal surface at constant velocity eventually stops. Why is this? Where has the kinetic energy of the moving ball gone? Although the ball loses kinetic energy, energy is conserved in the closed system of the ball and the ground. Due to air resistance and friction with the ground, some of the ball's kinetic energy is transformed into thermal energy in the surroundings, raising the temperature of the ground and the ball slightly. This shows that energy is never lost, although not all energy transformations are useful, or can be used to do work.

Radioactive decay

Throughout the Middle Ages, alchemists tried to discover how substances were made. They wanted to find a method of transforming a substance like lead into gold—a process they called transmutation. No alchemist ever achieved this change, but interest in the structure of matter continued. In 1901, experiments by Ernest Rutherford and Frederick Soddy showed that part of a sample of the element thorium had spontaneously transformed into the element radium. Rutherford jokingly warned Soddy, "Don't call it transmutation, they'll have our heads off as alchemists!"

The conversion of atoms of one element into atoms of another element occurs by radioactive decay. But not all radioactive decay results in transformation of atomic nuclei. In gamma decay, for example, an unstable nucleus becomes more stable by emitting gamma rays but the structure of the nucleus is unchanged.

⌘ CHAPTER LINKS

See Topic 3 in Chapter 11 on form for more about ways in which energy can be stored, and ways in which energy can be transferred from one store to another, such as by a swinging pendulum.

QUICK THINK

a) Describe the energy transformations in the following devices:
   • a diesel vehicle
   • a battery toy.

b) Is all the energy from the fuel or energy store transferred into a useful form? How does the law of conservation of energy apply in this situation?

Radioactive decay underlies many modern technologies, from nuclear power to the dating of rocks or archeological finds. In order to understand this phenomenon, we must first review the structure of the nucleus.

The nucleus of an atom contains protons and neutrons. The number of protons is called the atomic number, and the number of neutrons is calculated by subtracting the atomic number from the atomic mass. However, atoms have different forms called isotopes. Each isotope of the same element has a different number of neutrons, but the same number of protons. Some isotopes are stable, and some have unstable nuclei that undergo radioactive decay.

For example, carbon has an atomic number of 6. This means all atoms of carbon have 6 protons. The most common isotope of carbon, carbon-12, has an atomic mass of 12. It therefore has 12 – 6 = 6 neutrons. A different isotope of carbon, carbon-14, is created in the atmosphere when cosmic rays collide with nitrogen atoms. Carbon-14 has an atomic mass of 14 and has 14 – 6 = 8 neutrons in the nucleus (Figure 7.5).

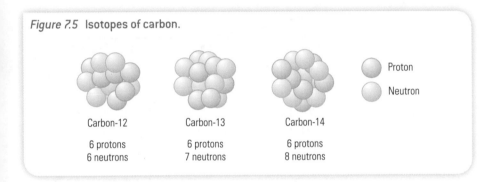

Figure 7.5 Isotopes of carbon.

Carbon-12
6 protons
6 neutrons

Carbon-13
6 protons
7 neutrons

Carbon-14
6 protons
8 neutrons

Proton

Neutron

Radioactive decay and half-life

Radioactive decay is random, so we cannot precisely predict when a particular nucleus will decay. Instead, scientists consider a large number of nuclei and predict the number that will have decayed after a certain amount of time. The time it takes for the radioactive count rate from a sample containing the isotope to fall to half its starting level is called the half-life. It is the time it takes for the number of nuclei of the isotope in a sample to halve.

Different isotopes have different rates of decay, so different isotopes have different half-lives.

Radioactive dating using isotopes

The carbon-14 isotope is much less common than carbon-12. However, it is incorporated into living tissue as plants undergo photosynthesis. As animals eat the plants, carbon-14 gets into their tissues. Carbon-14 is radioactive and decays into nitrogen 14, as a neutron breaks down into a proton and an electron. The proton stays in the nucleus and the electron leaves the atom. An electron antineutrino is also emitted.

QUICK THINK
Find out about the half-lives of different radioactive isotopes produced in nuclear power stations. What problems do these products cause?

The graph in Figure 7.6 shows how the amount of carbon-14 left in a sample decreases over a certain period of time. In one half-life, the amount of the original sample decreases by half (from 100 per cent to 50 per cent). During the second half-life, the amount of the original sample decreases by half (from 50 per cent to 25 per cent).

a) Use the graph to find the half-life of carbon-14.

b) Use the graph to find the percentage of carbon-14 left after 3 half-lives.

c) Is this the percentage you would expect?

Explain your answers.

Following the death of the animal or plant, the amount of radioactive carbon no longer increases (by photosynthesis or feeding) but the carbon-14 present continues to decay with a half-life of 5,700 years. The amount of carbon-14 in tissues follows a specific decay curve as shown in Figure 7.6. Scientists measure the amount of carbon-14 in an archeological sample of plant or animal tissue and, by comparing this to the ratio of carbon-14 to carbon-12 in living things today, the age of the sample can be determined. If there is half the amount of carbon-14 nuclei remaining in the sample compared to the amount in living tissue, then the sample must be 5,700 years old.

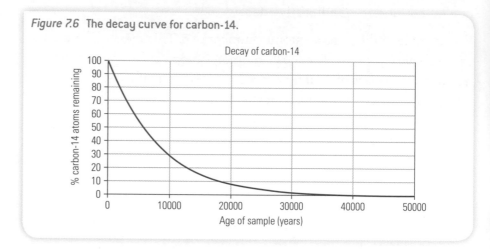

Figure 7.6 The decay curve for carbon-14.

As you can see in Figure 7.6, the amount of carbon-14 nuclei remaining falls to very low numbers after about 30,000 years. Carbon-14 dating is only accurate for objects up to about 60,000 years old.

Other radioactive isotopes, such as potassium-40, which has a half-life of 1.2×10^9 years, can be used for dating minerals and rocks from 100,000 to 4.3 billion years old. This is the age of the Earth.

 Activity 5   Beer foam decay curve

You cannot directly observe the decay of radioactive nuclei. You are going to use the rate of decay of foam on beer as a simulation or analogy. It is a useful way to learn about patterns of real phenomena that cannot be directly observed.

As beer or carbonated soft drinks are poured into a container, carbon dioxide bubbles rise to produce a frothy layer at the top of the container. Because the sides of the container are not completely smooth, they provide a site for the bubbles to form on and stick to. During this investigation, you will try to produce as much foam from the poured beer as possible. As individual bubbles burst, the beer foam will transform over time to its liquid state. The bubble-bursting occurs in a random way analogous to the random decay of individual radioactive nuclei. The transformation can be recorded as a decrease in the volume of the foam, and should create a decay curve similar to that seen in radioactive decay.

You will observe the decay of the height of the beer foam over time and create a graphical representation of the decay pattern.

Hypothesis

The volume of foam will decrease over time. As the volume of foam decreases, the rate of decay of the beer foam will also decrease. This is due to the fact that, over time, there is less foam remaining, and therefore less beer mass per unit of time that changes.

STEP 1 Collect together the following:

- 1 bottle/can of non-alcoholic beer
- large graduated cylinder (preferably 500 cm³)
- stopwatch
- ruler.

STEP 2 Read through the Steps 3–4 and construct a data table to record the data you will collect.

STEP 3 Pour the beer into the graduated cylinder to produce as much foam as possible. The best method is to pour the beer into the cylinder quickly, without tilting. Measure and record the height of foam produced. Be sure to subtract the height of liquid in the graduated cylinder. That is:

> height of foam = height from bottom of the foam (so the top of the liquid) to the top of the foam

TIP

Hints for drawing graphs.

- Choose an appropriate range and scale for each axis to make the graph fill as much of the page as possible, so it is easy to read.
- Always plot the independent variable (the variable changed during the experiment) on the x-axis, and the dependent variable (the measurements recorded) on the y-axis.
- Label the x and y axes (include units).
- Add a line of best fit. In order to help your reader, the trendline colour should be the same as the data points.
- A legend is necessary only if two different curves/lines are plotted on the same axes.
- Add a figure caption below the graph.

STEP 4 Start the stopwatch and record the height of the foam every 10 seconds until no foam remains on top of the liquid.

STEP 5 Plot a graph of your data using a spreadsheet or by hand.

Conclusion and evaluation

STEP 6 Using the data you collected, describe the relationship between the time taken and volume of foam. Is the gradient of the graph increasing, decreasing or remaining constant over time? What does this suggest about the rate of decay of beer foam?

Do you think that the rate of decay of foam would be the same if the brand of beer was changed? Explain your answer.

STEP 7 From your graph, find the half-life of your sample's foam between different pairs of time. Is the half-life constant?

Comment on the **accuracy** of your results, and how close your measurement of the half-life is to the value found by other groups.

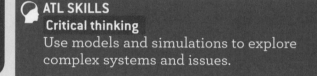

🌐 **GLOBAL CONTEXTS**
Scientific and technical innovation

🧠 **ATL SKILLS**
Critical thinking
Use models and simulations to explore complex systems and issues.

Radioactive dating is important in archeology and geology. The technique was first used in 1907, and since then has been improved so very small samples can be analysed with great accuracy. A variety of isotopes with different half-lives are used to determine the age of a wide range of rocks and artefacts. The age data helps scientists form and add to theories on evolution, estimates of the age of the Earth or of individual rock formations, anthropology and many other disciplines.

Reflection on Topic 2

Conservation laws are important in allowing us to predict changes in energy from one form to another. The conversion between different forms of energy can be seen in our daily lives. Atoms of one element can be changed into atoms of another element by radioactive decay.

- What forms of energy loss can occur during an energy transformation?

- What other ways can you think of to model radioactive decay without using radioactive materials?

Summary

Transformation is a critical concept to understand and apply to a variety of situations. Such situations include:

- transformation of energy in food chains

- moving objects

- change of state from solid to liquid to gas

- transformation of an atom into a different isotope or element when particles leave the nucleus.

In the particular case of transformation of energy in food chains, you have learned that energy is "lost" at each transfer (because organisms in lower trophic levels use much of the energy for their own functions or structures). However, energy is neither created nor destroyed; it is conserved and must be accounted for.

Models

INQUIRY QUESTIONS	
	TOPIC 1 Evaluating models in biology
	▪ **What can simple models tell us about complex phenomena?**
	▪ **What are the strengths and limitations of models for investigating natural phenomena?**
	▪ **What are the challenges involved in evaluating models?**
	TOPIC 2 Models of atomic structure
	▪ **How do models help us understand processes that may not be easily observed?**
	▪ **How does scientific knowledge change with time?**
	TOPIC 3 Making waves
	▪ **How do we use models to visualize something we cannot see, such as light and sound waves?**
	▪ **How can wave properties be predicted using models?**
	▪ **What affects the speed of a wave?**

GLOSSARY

Prediction an expected result of an upcoming action or event.

Validate to confirm or verify with evidence that a scientific model is true.

Validity of the method refers to whether the method allows for the collection of sufficient valid data to answer the question. This includes factors such as whether the measuring instrument measures what it is supposed to measure, the conditions of the experiment and the manipulation of variables (fair testing).

SKILLS

ATL

✓ Use models and simulations to explore complex systems and issues.

✓ Recognize unstated assumptions and bias.

✓ Gather and organize relevant information to formulate an argument.

✓ Make connections between various sources of information.

✓ Use and interpret a range of discipline-specific terms and symbols.

✓ Draw reasonable conclusions and generalizations.

✓ Make guesses, ask "what if" questions and generate testable hypotheses.

Sciences

✓ Design a method for testing a hypothesis, explaining how to manipulate the variables and how data will be collected.

✓ Evaluate the validity of a model based on the outcome of an investigation.

✓ Organize and present data in tables ready for processing.

✓ Draw sketches of observations from an experiment.

COMMAND TERMS

Draw represent by means of a labelled, accurate diagram or graph, using a pencil. A ruler (straight edge) should be used for straight lines. Diagrams should be drawn to scale. Graphs should have points correctly plotted (if appropriate) and joined in a straight line or smooth curve.

Sketch represent by means of a diagram or graph (labelled as appropriate). The sketch should give a general idea of the required shape or relationship, and should include relevant features.

Write down obtain the answer(s), usually by extracting information. Little or no calculation is required. Working does not need to be shown.

OTHER RELATED CONCEPTS

Energy   Evidence   Interaction   Movement   Patterns

Figure 8.1 Linus Pauling. He was later awarded a Nobel Prize in Chemistry and a Nobel Peace Prize

Introducing models

Proteins perform hugely varied and essential functions in living organisms. For example, collagen strengthens ligaments, insulin regulates blood glucose, enzymes such as pepsin break down proteins we eat and reuse the building blocks to build new proteins.

You may be surprised to find out that with such a vast array of different functions, the basic building blocks of proteins are remarkably similar. All protein molecules are made from long chains of smaller molecules called amino acids. These building blocks can be put together in millions of different ways to make millions of different proteins. The instructions for the order of amino acids in each protein are contained in our DNA code.

Proteins are intricately folded structures and each has a specific shape. In the early days of biochemistry, it was a great mystery how the chains of amino acids turned into the complex and very specific structures of proteins.

Then Linus Pauling (Figure 8.1)—one of the greatest scientists of the 20th century—got a bad cold and was forced to stay in bed for a couple of days in his London apartment in the winter of 1948.

Pauling knew the structure of amino acids, and he had seen X-ray crystallography images of proteins that hinted at their shape. But for almost ten years he'd struggled with the problem of how amino acids linked together to form a chain.

So, bored with being in bed and reading detective novels, he did what any scientist would do. He started to draw and build a model to help him with his problem.

One definition of a model is a representation used for testing scientific theories or ideas. The model can then be tested to see whether it accurately predicts something that can later be observed.

As the story goes, Pauling worked out an appropriate scale and arranged a chain of amino acids as in Figure 8.2. He then started to fold the paper in a way that would get as many oxygen atoms (shown as =O) in the carboxyl groups aligned with as many as possible of the hydrogen atoms (shown as –H) in the amine groups of another amino acid. The bonds formed between these groups give the protein strength and structure.

How? He rolled the paper into a kind of telescope shape that looked like a spiral and found that the hydrogen and oxygen atoms came together at regular intervals in what he called an alpha helix. The geometry of the model gave **predictions** about protein shape and structure that were not confirmed experimentally for a couple of years, but he had effectively solved this great mystery with little more than a pencil and a piece of paper—and the time he had from having a bad cold that kept him in bed!

Not all models are drawings of what they represent. Some models are ideas or descriptions of processes that may be difficult to observe or happen very slowly, very quickly or on too small a scale. Such models can be improved by testing their predictions on a small data set, and altering the values or ideas in the model until there is a better fit with the data.

Computer simulations are also models. Again, they can be used to explain or predict processes that are difficult to observe. They can also help us to understand the dynamics of multiple underlying phenomena of a complex system such as the weather. Simulations can be repeated again and again using the same initial conditions to **validate** the mathematical equations in the model (that is, to be certain the equations are an accurate representation of the scenario being modelled).

Models are powerful tools in the understanding of difficult scientific ideas and phenomena. They evolve and become more complex as scientists use them and refine or improve them.

Figure 8.2 The drawing Linus Pauling made of his model of the alpha helix, showing how the paper should be folded or rolled so that the side group extending from one amino acid makes a bond with the side group of another amino acid, in a regular, repeating pattern

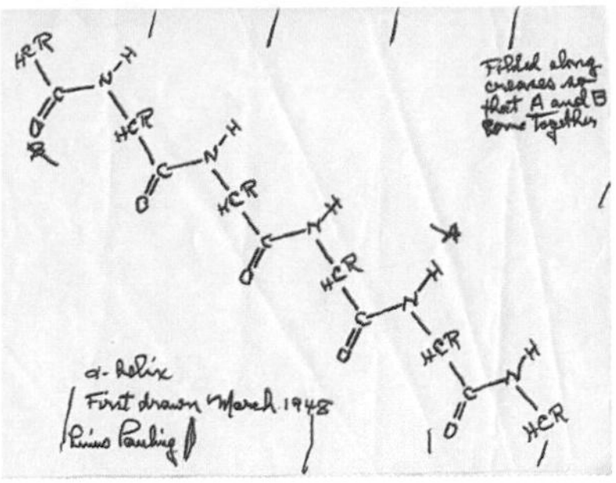

TOPIC 1

Evaluating models in biology

Scientists use models for a number of reasons. For example, studying a biological phenomenon in its natural context might be too difficult if the time period over which it occurs is too long, or the scale of the phenomenon is too small or too large. The complex interactions within a system might make it difficult to isolate and control variables in order to carry out a fair test.

Useful models should allow predictions and should offer a means to explain observations. They commonly include mechanisms that parallel the phenomenon they are meant to model.

Evolution is the change in observable characteristics in a population with time. A change in observable characteristics in a population can be observed over a relatively short time period in some cases (such as the development of antibiotic resistance, selective breeding). However, in most cases, evolutionary change is much too slow for humans to observe easily. Modelling provides the opportunity to study adaptation and the evolution of forms and functions to suit environmental pressures.

Some marine organisms dive for long periods of time and to great depths in order to find food. For example, the Weddell seal can dive for up to an hour when hunting (Figure 8.3). The challenge is to conserve oxygen used in respiration by minimizing energy expenditure when swimming.

Figure 8.3 Weddell seals

In Charles Darwin's theory of evolution by natural selection, evolution occurs when environmental selection pressures lead to differences in reproduction rates of some varieties over others, usually because of the death of poorly adapted individuals before they reproduce.

 Activity 1   Modelling evolution by natural selection

How did seals evolve a streamlined body shape? This investigation is a model that simulates evolution by repeatedly applying a selection pressure so that only some of each generation of models survives and reproduces.

You will investigate what body shape adaptations give a diving organism an advantage for moving through water. By reducing "drag", a diving organism can go fast without having to waste too much energy.

STEP 1   Gather together the following apparatus:

- stopwatch
- tall graduated cylinder
- modelling clay.

STEP 2   Make 10 or more model animals using modelling clay, or some other material that can be easily formed into different shapes. Drop each of them into a measuring cylinder of water. Measure and record how long each takes to reach the bottom.

STEP 3   Discard the half of the models that were slowest. Formulate a hypothesis as to what were the features of the surviving models that allowed them to sink faster.

STEP 4   Pair up the fastest models and make intermediate shapes, to represent their offspring. The offspring should have some of the features of each parent.

STEP 5 Test the new generation of model animals and repeat the elimination of the slowest and the simulated breeding of the fastest.

STEP 6 Does one shape gradually emerge? Describe its features. **Sketch** drawings of your intermediate models. Compare your speediest model to the speediest models of other groups.

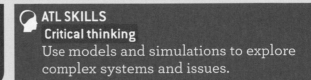

GLOBAL CONTEXTS
Orientation in space and time

ATL SKILLS
Critical thinking
Use models and simulations to explore complex systems and issues.

Evaluation is a very important skill to develop in the sciences. In many ways, it is like being asked for your opinion, although it requires you to reflect deeply on the data, the topic or the concept or the question before you tell everyone what you think about it and why.

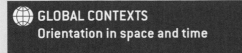 Activity 2   Modelling movement across a cell membrane

In living things, an important function of cellular membranes is to regulate the movement of materials in and out of cells. This can occur spontaneously, when particles move from an area of high concentration to an area of low concentration.

It is difficult to carry out observations on movement across a single cell membrane but artificial membranes can be used as models of cell membranes. When an individual's kidneys no longer function properly, his or her blood must be artificially filtered by a process known as dialysis. Tubing made for use in kidney dialysis machines can be used to model a cell membrane. This tubing is semipermeable. This means that it contains small pores that allow fluid and small solute molecules to pass through, but not blood cells and proteins. Osmosis is the diffusion of water across a semi-permeable membrane.

STEP 1 Gather together the following materials:

- eye protection
- Visking tubing
- dialysis clamps
- flat carbonated cola
- a scale

- a beaker
- Benedict's solution
- beaker of water heated to 100°C (water bath for Benedict's test)

- test tube
- pH indicator.

[SAFETY] Wear eye protection. Do not drink or taste the cola. Avoid skin contact with Benedict's solution, especially when heated.

STEP 2 Obtain some cola soft drink and allow it to go flat—in other words remove the carbonation—by leaving it open overnight.

STEP 3 Cut a section of tubing and soak it. Knot the tubing at the bottom or tie it very tightly with some string. Pour the cola through the open end and then knot or clamp the remaining end. Do not fill the bag with cola.

STEP 4 Measure and record the mass of the Visking tubing plus cola and then place it into a beaker of distilled water.

Cola contains fructose, phosphoric acid and caramel, a brown colouring substance.

STEP 5 Predict which of these substances—if any—will diffuse out of the Visking tubing, giving reasons for your predictions.

STEP 6 Predict whether the mass of the Visking tubing after it has been in the water for one hour will increase or decrease.

STEP 7 After one hour, remove the Visking tubing from the beaker, dry off the water and measure and record the mass of the tubing.

STEP 8 Look carefully at the colour of the water in the beaker to see whether it is clear or if it has a brown cola colour. Using a digital camera such as a computer or tablet camera, you could take photos at the start and at the end of the experiment to facilitate the colour comparison.

The presence of fructose in the water in the beaker can be detected with the Benedict's test.

STEP 9 Add 2 cm³ of the water from the beaker to a test tube and add an equal volume of Benedict's solution. Shake well to mix. Heat the test tube in the water bath for about 5 minutes, or until the colour of the mixture changes. The original pale blue colour means no fructose is present. A red colour means fructose is present.

The presence of phosphoric acid can be detected using a narrow range pH indicator.

STEP 10 Test the pH of the water in the beaker.

a) Interpret what your results show.
b) Suggest what features of a real cell membrane are missing from this model.

GLOBAL CONTEXTS
Scientific and technical innovation

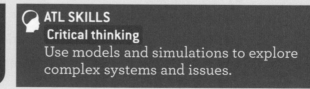
ATL SKILLS
Critical thinking
Use models and simulations to explore complex systems and issues.

In addition to the dialysis tubing used in Activity 2, biological tissues can be used as models of cell membranes. In this investigation, grape cubes are used to model osmosis in plant tissues.

STEP 1   Gather together the following material and apparatus:

- knife or scalpel
- cutting board
- fresh grapes
- honey
- three small beakers.

[SAFETY] Do not eat or taste the grapes or honey. Take great care when using the knife or scalpel as your fingers will be very close to the blade.

STEP 2   After you have read the rest of the instructions, construct a data table to record the measurements you will make.

STEP 3   Using a knife and a cutting board cut all of the skin, and some of the flesh from three grapes to produce three grape cubes. Use the ruler to check that the cubes are all the same size with sides of at least 10 mm. Be careful in your measurements as changes to the grapes during this investigation will be small. Record the initial measurements.

STEP 4   Place one grape cube in each of three small beakers.

STEP 5   Cover the grape in one beaker with liquid honey, the grape in the second beaker with water and the grape in the third beaker with a 50:50 mixture of liquid honey and water.

STEP 6   Once you have completed setting the experiment up, predict the outcome giving reasons.

STEP 7   Leave the grape cubes in the liquids for 24 hours, then take each of them out and place on a paper towel.

STEP 8   Turn the cubes over a few times to remove the honey and dry off the water. Use the ruler to measure the dimensions of each cube. Press each one with a finger to feel whether it is hard or soft.

a) Determine whether the dimensions of the side of the cube increased or decreased in diameter, or remained unchanged.
b) State whether each cube is turgid (hard) or flaccid (soft).
c) Explain any changes that have occurred.
d) In this investigation, plant tissue is being used to model an individual plant cell. Discuss the limits of this model by comparing tissues to cells.

 **GLOBAL CONTEXTS**
Scientific and technical innovation

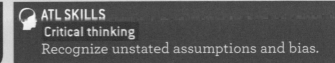 **ATL SKILLS**
Critical thinking
Recognize unstated assumptions and bias.

In this investigation, an egg with its shell removed will be used to model an animal cell. The white-coloured membrane in the scanning electron micrograph is the chorion. The brown broken egg shell is at the top. The chorion and egg shell control the passage of gases in and out of the egg. Once the egg shell is removed, the rubbery white surface that remains is the chorion.

STEP 1 Gather together the following materials and apparatus:

- eye protection
- 200 cm³ beaker
- three raw chicken eggs
- weak acid such as vinegar
- two sodium chloride solutions (10 per cent and 20 per cent)
- top pan balance.

[SAFETY] Wear eye protection. Wash hands after handling raw eggs.

STEP 2 Prepare the three eggs. The shells of chicken eggs can be removed by leaving them for 24 hours in a beaker containing a weak acid such as vinegar. You may need to turn the eggs over to ensure the shell on the top portion of the egg that floats above the solution is also removed. Alternatively, a smaller weighted beaker can be used to hold the egg submerged.

STEP 3 Place each of the eggs on a paper towel and blot them dry by gently patting them. Determine the mass of each egg and record the mass in a table of your own design.

STEP 4 Put each egg in a 200 cm³ beaker. Prepare two sodium chloride solutions (10 per cent and 20 per cent) and some distilled water (0 per cent sodium chloride). Cover each egg with 150 cm³ of one of the solutions. Leave the egg in the solution for 24 hours.

STEP 5 At the end of this period, pour the solution off the egg. Pat the egg dry on a paper towel. Record the mass of each egg in your table. Determine the change in mass of each egg.

TIP

$$\text{percentage change} = \frac{\text{change}}{\text{original}} \times 100$$

In this experiment, the change in mass is calculated by subtracting the original mass from the final mass.

STEP 6 Calculate the percentage change in mass to make comparison between the three eggs easy.

STEP 7 In this experiment, the shell-less egg is used to model an individual animal cell. The exposed rubbery white surface of the egg is the multicellular chorion. Discuss the limits of this model.

 **GLOBAL CONTEXTS**
Scientific and technical innovation

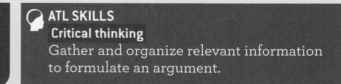 **ATL SKILLS**
Critical thinking
Gather and organize relevant information to formulate an argument.

Reflection on Topic 1

- What are the strengths and limitations of models for investigating natural phenomena?
- What can simple models tell us about complex phenomena?
- What are the challenges involved in evaluating models?

Models of atomic structure

There are many examples of different models in chemistry, such as the periodic table, drawings or 3D models of the shapes of molecules. In this topic you will explore how models of atomic structure have changed over time and how well they explain processes that may not be observable with the naked eye.

The ancient Greeks were the first to propose that atoms existed. The word atom is derived from the Greek word *atomos*, which means uncuttable or indivisible.

The ancient Greeks proposed that there were four types of atom with different shapes—earth, wind, water and fire. This was a very early theory of the nature of matter. Theories are integrated, comprehensive models of how the Universe, or parts of it, work. A theory can incorporate facts, laws and tested hypotheses. The scientist credited with first developing the modern theory of atomic structure was John Dalton in 1808.

Models like Dalton's are created to explain processes that may not be easily observed. When tested against experiments, models may be found to have flaws. When new experiments are carried out, evidence may be collected that cannot be explained through the use of the current model. In such a case, the new observations and data either lead to modifications of the existing model or a completely new model. The new model allows for better explanation of the new observations.

Based on new observations and experiments, Dalton's model of the atom was superseded by Thomson's plum pudding model of the atom (Figure 8.4). A more modern analogy would be a chocolate chip cookie model, in which electrons are the chocolate chips surrounded by positively charged cookie dough. Thompson's model had no nucleus, but after the gold foil experiment designed by Ernest Rutherford, the model of the atom included a small, dense nucleus.

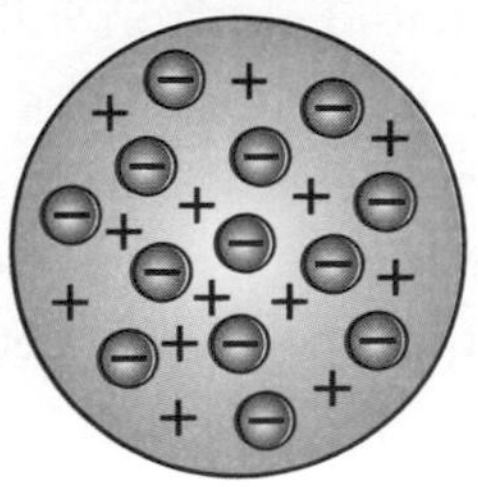

CHAPTER LINKS
In Chapter 9 on interaction you will construct molecular models.

WEB LINKS
Find out more about Dalton's atomic theory by searching for Dalton and Frostburg.

Figure 8.4 Thompson's plum pudding model of the atom

QUICK THINK

When you read about Dalton's atomic theory, you will find two assumptions that were changed as a result of new evidence that was discovered after Dalton. Comment on that evidence and how it meant that Dalton's model of the atom needed to be changed.

WEB LINKS

Rutherford's model proposed that most of the atom is empty space. How do solid substances exist if most of an atom is empty space? Consider this while watching the short clip "The space between atoms" found at www.youtube.com.

CHAPTER LINKS

In Chapter 14 on patterns you will look at how Bohr's model explains the pattern in reactions of elements in the periodic table.

WEB LINKS

An animation of Bohr's model of the atom can be found by searching for SBCC and Bohr's model.

The model of the atom that most closely aligns with the modern atomic theory was proposed by Niels Bohr in 1913 (Figure 8.5). Bohr's model of the atom differed from Rutherford's as Bohr proposed that electrons could only exist at certain distances from the nucleus. The electrons move around the nucleus in the same way that the planets orbit the Sun.

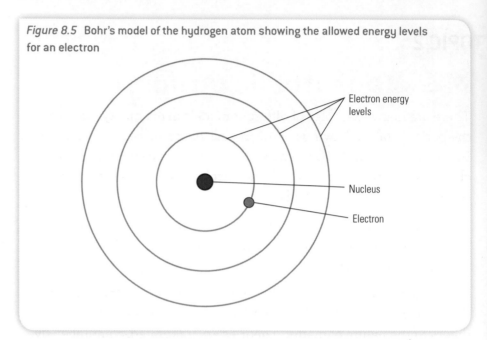

Figure 8.5 Bohr's model of the hydrogen atom showing the allowed energy levels for an electron

Electron energy levels

Nucleus

Electron

The modern model of the atom is based on a quite complicated theory of quantum mechanics developed by Erwin Schrödinger. As in Bohr's model, electrons can only have fixed energy levels. But in Schrödinger's model, the exact position of an electron cannot be determined.

One piece of evidence that confirms Bohr's model of the atom is the existence of line spectra. In Bohr's model, each energy level represents a specific energy state for the electron. Electrons in orbits nearest the nucleus have the smallest amount of energy and electrons in orbits furthest from the nucleus have the highest amount of energy. When an additional amount of energy is supplied, for example by heating, electrons can become excited and jump to a higher energy level. However, this move is not permanent. Very soon, an excited electron will move back to an inner orbit and emit energy equal to the difference of the energies of the two orbits. Figure 8.6 shows the emission of a photon (packet) of light as an electron in an atom moves from an excited state back to its lowest energy state, the orbit nearest the nucleus.

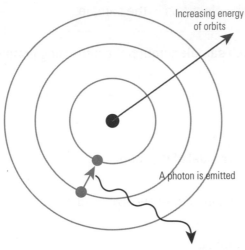

Figure 8.6 Bohr's explanation of emission of electromagnetic radiation from an excited atom.

Increasing energy of orbits

A photon is emitted

An incandescent light bulb or other very hot dense object emits a continuous spectrum of light. This means that all colours of light are visible and there are no gaps between the wavelengths of light – the spectrum is continuous. However, light emitted by a hot transparent gas is a line spectrum, in which only certain frequencies of light are emitted. Specific frequencies correspond to distinct colours of light. Continuous and line spectra are compared in Figure 8.7.

Figure 8.7 Continuous spectrum and atomic emission spectrum (line spectrum).

Continuous spectrum

Wavelength →

Emission line spectrum

Wavelength →

⊂⊃ CHAPTER LINKS
Topic 3 in Chapter 6 on evidence has more about the electromagnetic spectrum and the different wavelengths of light, and the relationship between wavelength and frequency.

 Activity 5　Bohr model of the atom and line spectra

In this activity, you will observe both continuous and line spectra and use knowledge of atomic structure to explain line spectra.

STEP 1　Obtain a diffracting grating. Diffracting gratings work by splitting light emitted from a source into its component wavelengths.

STEP 2　Look through the grating at an incandescent light source (filament bulb). **Sketch** what you observe.

STEP 3 Your teacher will connect a source of high voltage to a low-pressure sample of an element, such as a gas discharge tube. This energy will excite electrons in the sample.

STEP 4 Observe the spectrum of light emitted by each sample through the diffracting grating. **Sketch** what you observe.

STEP 5 Repeat steps 3 and 4 for each sample.

STEP 6 Describe how your observations reflect Bohr's model of the atom.

STEP 7 Explain why the samples only showed a colour when a current was passed through the discharge tube.

STEP 8 Explain why different substances show different spectra.

🌐 **GLOBAL CONTEXTS**
Scientific and technical innovation

🧠 **ATL SKILLS**
Information literacy
Make connections between various sources of information.

QUICK THINK

How could scientists use line spectra to study the stars?

 **WEB LINKS**
You can examine line spectra more closely by searching for Line Spectra at www.learnerstv.com.

Subatomic particles

There are three main types of particle that are found in an atom—the electron, proton and neutron. Protons and neutrons are often referred to as nucleons as they are found in the nucleus. The nucleus is the central region of the atom where most of its mass and all of its positive charge is concentrated.

The elements in the periodic table are arranged by increasing atomic number. An element's atomic number (Z) indicates how many protons are in each atom of that element. An element's mass number (A) indicates the number of protons and neutrons that are found in each atom of that element.

In neutral atoms, the number of protons equals the number of electrons. Ions are atoms or molecules that do not have an equal number of protons or electrons. Ions can either have a positive or negative charge caused by the loss or gain of electrons.

 **Activity 6** Mapping the inside of an atom

Why is it that the number of protons in an atom of an element cannot change?

Deduce the missing values in the table below. You will need to use a periodic table to identify the element if this is not shown. To find the mass number, round the atomic weight from the periodic table to the nearest whole number. Remember to add a charge to the symbol if you deduce it is an ion.

Symbol for element or ion	Atomic number	Mass number	Number of protons	Number of electrons	Number of neutrons
Bi	83	209			
Pt		195			
	77				
Ra				88	138
Xe		131			
V					
Al	13	27			
	99				
		190			
Sn	50	119			
Ag		108			
	53	127		53	
Ar				18	22
W		184			
Cs					
	65				
		98			
Cu^{2+}					
S^{2-}					
Cl$^-$					
Hg^{2+}					
	22			20	
	35			36	
	16			18	
Au$^+$					
F$^-$					
Zn^{2+}					
	29			27	
O^{2-}					

🌐 GLOBAL CONTEXTS
Scientific and technical innovation

ATL SKILLS
Communication
Use and interpret a range of discipline-specific terms and symbols.

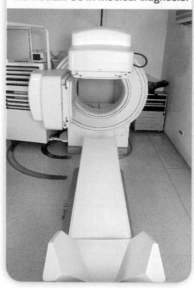

Figure 8.8 The largest use of radioactive isotopes is in the field of medicine. This scanner uses technetium-99 in medical diagnosis.

Isotopes

By definition, all atoms of the same element must have the same number of protons, but they need not all have the same number of neutrons. Atoms that have the same number of protons but different numbers of neutrons are called isotopes. Isotopes of the same element have the same atomic number but different atomic masses. They have the same chemical properties but they have slightly different physical properties. For example, deuterium is an isotope of hydrogen that has one extra neutron in the nucleus. Heavy water is water that contains a greater than usual proportion of the deuterium isotope 2H in place of 1H. The density of heavy water is about 10% greater than that of ordinary water.

There are about 2,000 known isotopes. Most of these are isotopes of radioactive elements, which can be extremely useful (Figure 8.8).

CHAPTER LINKS
In Topic 2 in Chapter 7 on transformation, decay of the foam on beer is used as a model of radioactive decay.

Activity 7   Isotopes—Beanium lab

In this activity, you will use dried beans as a model for atoms. You will measure the masses of individual beans and the relative proportions of different beans in a sample to show how an element's relative atomic mass is calculated.

STEP 1 Read through the rest of the instructions and design a table to record the data that will be collected.

STEP 2 Your teacher will give you a sample of the element "beanium". You will notice that different atoms of beanium appear different; these are different isotopes of beanium.

STEP 3 Sort the atoms of beanium into isotope groups. **Write down** the total number of atoms of each isotope and the number of isotopes.

STEP 4 Measure the total mass of each of the isotope groups.

STEP 5 Calculate the mean mass of a single atom of each isotope; this value is called the isotopic mass.

STEP 6 Calculate the isotopic abundance (percentage) for each isotope.

STEP 7 Calculate the relative atomic mass for beanium based on the isotopic abundances and the isotopic masses.

relative atomic mass =

$$\left(\text{isotopic mass} \times \frac{\text{isotopic abundance}}{100}\right) + \left(\text{isotopic mass} \times \frac{\text{isotopic abundance}}{100}\right) + \dots$$

STEP 8 For practice, calculate the relative atomic mass for the following isotopes:

a) 98.89% carbon-12, 1.11% carbon-13

b) 99.63% nitrogen-14, 0.37% nitrogen-15

c) 78.70% magnesium-24, 10.13% magnesium-25, 11.17% magnesium-26

d) 92.21% silicon-28, 4.70% silicon-29, 3.09% silicon-30

e) 15.8% molybdenum-92, 9.10% molybdenum-94, 15.7% molybdenum-95, 16.5% molybdenum-96, 9.5% molybdenum-97, 23.8% molybdenum-98, 9.6% molybdenum-100.

 **GLOBAL CONTEXTS**
Scientific and technical innovation

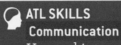 **ATL SKILLS**
Communication
Use and interpret a range of discipline-specific terms and symbols.

Reflection on Topic 2

Scientists today are still refining the model of the atom. Their discoveries could change our current thinking about the atom.

- How important was the use of models for scientists who were investigating atomic structure?

- Why is it suitable for us to base our study of the model of the atom on Bohr's model of the atom, although we know that a more complex model exists?

- What are the challenges involved in evaluating models?

TOPIC 3

Making waves

In science, there are many opportunities to learn from simulations and models when a phenomenon is not directly observable.

A scientific model can be an idea, a description or a substitute for what it represents (eg the Visking tubing model in Topic 2, above). Ideas about related processes or structures can be represented in the form of models, and the model can then be tested to see if it is valid.

Scale models are a different type of model. For example, a scale model of the Solar System can be made using balls of different sizes in concentric circles at various distances from a central Sun. We can easily observe the properties of systems using such models.

Mathematical models derive predictions of future events based on past patterns. Climate models have given us a glimpse of the predicted changes to the Earth's mean global surface temperature as global

QUICK THINK

With reference to the beer foam activity from Topic 2 in Chapter 7, discuss the strengths and weaknesses of using a model to make predictions about a scientific phenomenon.

carbon emissions continue to rise. Climate models represent energy and mass transfer processes in the atmosphere and ocean. They are continually evaluated and revised as more data is collected.

Types of wave

A wave transfers energy from one place to another. There are two types or forms of wave: transverse and longitudinal.

Transverse waves

In these waves, particles move perpendicularly to the direction in which the wave is travelling (Figure 8.9). Some examples of transverse waves are water waves (in which a mass of water moves up and down, transmitting energy outwards) and light waves (in which a charged particle oscillates in an electromagnetic field, transmitting energy through a vacuum).

Longitudinal waves

In these waves, the particles move in the same direction as the direction in which the wave is travelling (Figure 8.9). An example is sound waves, where a material is compressed and decompressed as the vibration travels through. Longitudinal waves require a medium in order to transmit energy.

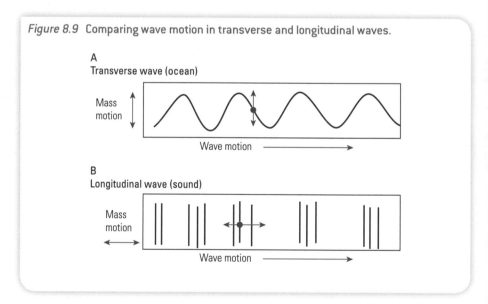

Figure 8.9 Comparing wave motion in transverse and longitudinal waves.

Earthquakes can cause massive destruction and endanger many lives. Faults are fractures in the Earth's crust that can shift suddenly, releasing stored strain energy that spreads outwards during an earthquake. The energy is released from a location called the focus where the crust is moving. The energy spreads out from the focus as seismic waves passing through solid rock.

The 2004 Sumatra earthquake ruptured a fault line 1,500 kilometres long. The fault line was caused by two tectonic plates sliding over one another, deep in the Earth's crust beneath the Indian Ocean. The ocean

CHAPTER LINKS
Topic 3 in Chapter 6 on evidence has more information about wavelength and frequency of a wave.

QUICK THINK

Models are being developed that attempt to predict when earthquakes will occur, at the locations where strain is known to be building up, and to limit the damage caused by earthquakes. Investigate how scientists attempt to develop these models.

WEB LINKS
Around the Pacific Ocean is a boundary where there are frequent earthquakes and many active volcanoes. To learn more, search for "ring of fire" at education. nationalgeographic.com/ education.

To learn about models of earthquake prediction, search for prediction at earthquake.usgs.gov.

floor above was suddenly uplifted by about 2 metres producing a series of large tsunami waves that travelled through the ocean to Indonesia and parts of Thailand, Maldives, India, Sri Lanka and Somalia. This earthquake was the third most powerful in the past 100 years according to the United States Geological Survey (earthquake.usgs.gov/earthquakes/world/10_largest_world.php).

The resulting tsunami killed over 230,000 people in 14 countries. In the aftermath of this unprecedented natural disaster, a tsunami warning system was put in place in the Indian Ocean, similar to the one already operational in the Pacific Ocean where earthquakes are more frequent.

An earthquake generates longitudinal and transverse seismic waves, as well as other types of seismic wave. In order to understand earthquakes, you are going to model these two types of wave using a Slinky spring toy.

 Activity 8   **Modelling transverse waves**

In this activity, you will observe the properties of a transverse wave using a metal spring as a model. The coils of the spring represent particles in solid rock. Moving the first coil of the spring represents an energy source, such as an earthquake. Each coil of the spring moves because the coil next to it transfers energy to it.

If you do not have a Slinky spring, you can use a long, flexible rope—the longer the rope the better, at least 3 metres. You could do this in a long, wide corridor or the school hall.

STEP 1 Collect the following:
- Slinky spring
- metre rule
- tape
- stopwatch.

TIP

Do not overstretch the Slinky spring because it will not return to its original shape.

STEP 2 Working with a partner on a smooth floor, table or bench, stretch the Slinky between you.

Send a single wave pulse towards your partner by quickly moving your hand to one side and then back to its original position, as shown opposite. Your partner must hold the end of the Slinky in a fixed position.

Observe the movement of the spring. How do the coils in the spring move compared to the direction in which the energy is transmitted?

Observe what happens to the wave once it reaches your partner. The returning wave is called the reflected wave. **Sketch** what you observe. This is called a fixed-end reflection.

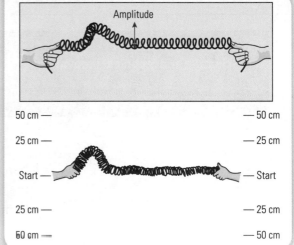

STEP 3 Have your partner let go of the Slinky and just leave the end on the floor. Send another wave pulse down the Slinky by moving your hand to one side and then back to its original position as in Step 2. Observe what happens to the wave once it reaches the end of the Slinky. How does the returning wave differ from the one seen in Step 2?

This is called a loose-end reflection.

STEP 4 Using a piece of tape, mark where the beginning of the Slinky lies on the floor, and measure 25 cm and 50 cm to the left of the starting point. Use pieces of tape to mark those locations. Repeat to the right of the starting point and also on the other end of the Slinky.

STEP 5 Your partner should now hold a stopwatch in one hand, and the end of the Slinky in the other. Create a wave with amplitude 25 cm, and use the stopwatch to time how long the wave takes to travel to the end. Record this time.

STEP 6 Repeat Step 5 but create a wave with 50 cm amplitude. You may need to practise this. Record the time for this wave to travel to the other end. Does the time of wave travel change?

STEP 7 Do at least three trials for each amplitude wave and find the mean travel time. Record all your data. Is there a relationship between amplitude and speed?

GLOBAL CONTEXTS
Scientific and technical innovation

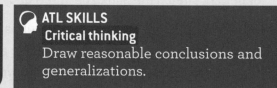
ATL SKILLS
Critical thinking
Draw reasonable conclusions and generalizations.

When a seismic wave hits a boundary between different rock types, some of the wave's energy will be reflected. Geophysicists can use information from reflected seismic waves to produce a model of the structure of rocks beneath the surface.

You have seen that a wave on a Slinky can be reflected back towards the source. If two waves travelling in opposite directions meet up, these opposing waves create what is called wave interference.

Activity 9   Modelling wave interference

In this activity, you will model interference of two travelling waves using two single wave pulses on a Slinky.

STEP 1 Collect the following:
- Slinky spring
- metre rule
- tape.

STEP 2 **Two waves of equal amplitude**
One partner creates a single wave pulse of amplitude 25 cm, while at the same time the person at the other end creates a single wave pulse of amplitude 25 cm on the same side (see diagram opposite). Observe what occurs when the waves meet. **Draw** a series of diagrams of the interference you observe.

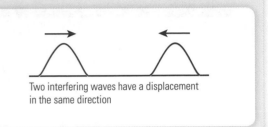
Two interfering waves have a displacement in the same direction

STEP 3 Repeat Step 2 but with both waves displaced to the other side of the starting point.

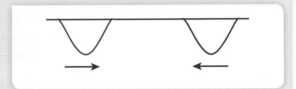

STEP 4 **Two waves of different amplitude**
One partner creates a single pulse of amplitude 25 cm, while at the same time the other person creates a single pulse of amplitude 50 cm wave to the same side. Observe and **draw** what occurs when the waves meet.

STEP 5 **Two waves of opposite displacement**
One partner creates a single pulse of amplitude 25 cm, while at the same time the other person creates a single pulse of amplitude 25 cm on the opposite side. Observe and **draw** what occurs when the waves meet.

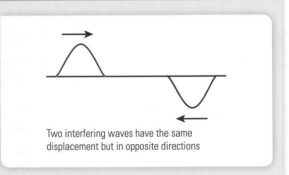
Two interfering waves have the same displacement but in opposite directions

STEP 6 **Two opposite waves—different amplitudes**
One partner creates a single pulse of amplitude 25 cm, while at the same time the other person creates a single pulse of amplitude 50 cm on the opposite side. Observe and **draw** what occurs when the waves meet.

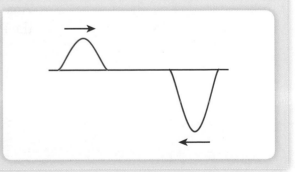

**STEP 7** From your observations, describe how transverse waves behave when they interfere with a:

a) wave of the same amplitude (Steps 2 and 3)
b) wave of greater amplitude (Step 4)
c) wave of the same amplitude but opposite displacement (Step 5)
d) wave with greater amplitude and opposite displacement (Step 6).

🌐 **GLOBAL CONTEXTS**
Scientific and technical innovation

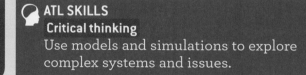💭 **ATL SKILLS**
Critical thinking
Use models and simulations to explore complex systems and issues.

The speed of seismic waves depends on the composition of the rocks along the wave's path, and how deeply they have travelled in the Earth. Compressional waves and transverse waves travel at different speeds through the same rock type. Compressional waves travel faster, and arrive at seismometer stations first. This is why they are called primary waves, or P-waves. Transverse waves travel slower, and arrive at seismometer stations second. This is why they are called secondary waves, or S-waves.

If scientists know the speeds of P-waves and S-waves at different depths, they can use the difference in time between arrival of the first P-waves and the first S-waves to estimate how far away the earthquake occurred. This is important if a large earthquake under the sea is expected to cause a tsunami—a tsunami alert can be sent to specific countries to warn people to move away from the coast.

This model for finding the location of an earthquake makes predictions that can be verified. For example, data for the same earthquake can be collected from several seismometer stations and the predictions compared. The accuracy of the model depends on knowing the speeds of the P-waves and S-waves.

Another kind of model used by seismologists is for the internal structure of the Earth. The travel times of seismic waves can be used to infer information about geological structures beneath the surface. Oil companies use methods like this to search for suitable drill sites. Again, the accuracy of the model depends on knowing the speeds of the P-waves and S-waves.

Activity 10 Exploring wave speed

In this activity, you will design an investigation to test how one property of a transverse or longitudinal wave will affect the speed of the wave.

You should select a range of five values for the independent variable, and test each value three times. By repeating each measurement three times, the mean value will be more accurate. Consider changing one of the following for the independent variable:

- amplitude
- number of coils in a wave
- wavelength
- diameter of Slinky
- medium of wave travel (rope/spring/string).

All other variables that could have an influence on the value of the dependent variable should be kept constant. Brainstorm this with a partner before creating your hypothesis and procedure.

STEP 1 Choose one variable to investigate. Explain why you think this variable will affect the speed of a wave.

STEP 2 Describe a detailed set of procedures for how you will test this hypothesis. Describe how you will alter the variable and collect data.

State the:

a) independent variable (an aspect of the wave you want to change)
b) dependent variable (what aspect of the resulting wave you will measure)
c) control variables (all other variables that must stay the same during trials)
d) materials (the materials and equipment you will need to test your hypothesis)
e) detailed method (so that anyone can repeat your experiment).

STEP 3 Construct a data table to record the data that you will collect. Include the units for the measurements.

STEP 4
a) Discuss what you can conclude from the data collected. How does your graph support the conclusion?
b) Does your conclusion agree with your hypothesis?
c) Evaluate the **validity of the method**, the equipment used and the accuracy of the data. Are there any outliers in your data (anomalies that do not fit the pattern)?
d) Suggest how you would improve the procedure if you were to carry it out again.

 **GLOBAL CONTEXTS**
Scientific and technical innovation

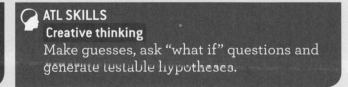 **ATL SKILLS**
Creative thinking
Make guesses, ask "what if" questions and generate testable hypotheses.

Models of interference have been used to develop the science behind noise-cancelling technology. The headphone has a hidden microphone to detect external background sound. The headphone speakers then create waves that are the inverse of the background sound. The inverse wave has the same amplitude and frequency but opposite displacement, causing interference that cancels out the noise coming from the environment. In Figure 8.10, the red wave shows the external sound wave that the listener does not want to hear, and the green wave shows the inverse wave created by the headphone speaker to interfere with the noise. In this way, the external sound wave is cancelled out and the result is no sound. The listener will be able to listen to the sounds he or she wants to hear coming through the headphones without distraction.

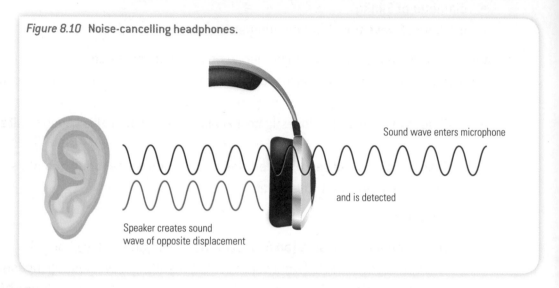

Figure 8.10 Noise-cancelling headphones.

Sound wave enters microphone

and is detected

Speaker creates sound wave of opposite displacement

CHAPTER LINKS
You will be looking at the concepts of constructive and destructive interference again, when studying standing waves in Chapter 14 on pattern.

This cancellation of the waves is referred to as destructive interference. A similar phenomenon of constructive interference occurs when two waves of the same amplitude and frequency coincide so that they reinforce each other, giving a wave of twice the amplitude.

Reflection on Topic 3

Wave properties can be predicted using models. For example, as travelling waves interfere with one another, predictable patterns emerge for the position and amplitude of the resultant waves.

By observing changes in wave behaviour as seismic waves travel through the Earth's interior, scientists can also learn about the structure and composition of the Earth.

- Discuss how the properties of water waves can be described using our models of constructive and destructive interference.

- Discuss the relationship between speed, frequency and wavelength of a wave. How can this equation help model the damage inflicted by tsunamis?

Summary

You have learned about the power of using models as you worked through this chapter. Models help to study the motion of molecules in and out of cells, subatomic structure and how to predict earthquakes. The activity on modelling evolution by natural selection shows that models are very useful to deduce relationships where long timescales make other methods impractical. Other situations where modelling proves to be a practical approach to representing relationships are when factors such as scale (from the very small to the very large) or volume of data make studying phenomena difficult.

You should have gained a sense that models are only ever as good as their ability to predict the unobservable and the complex, and are constantly undergoing the process of refinement to make them more accurate as more data is tested against the model.

Interaction

**INQUIRY
QUESTIONS**

TOPIC 1 Interactions between organisms

- **What factors influence leaf damage caused by herbivorous insects?**
- **How do conifers affect germination of seeds of other plants?**
- **What is the most effective strategy for removing plaque from teeth?**

TOPIC 2 Chemical bonding

- **What role does chemistry play in material science?**
- **What are the similarities and differences between ionic and covalent compounds?**

TOPIC 3 Electromagnetism

- **How do electric currents and magnetic fields interact?**
- **How is an electromagnet created?**
- **What can affect the strength of an electromagnet?**

SKILLS

ATL

✓ Make guesses, ask "what if" questions and generate testable hypotheses.

✓ Collect and analyse data to identify solutions and make informed decisions.

✓ Draw reasonable conclusions and generalizations.

✓ Make connections between various sources of information.

✓ Present information in a variety of formats and platforms.

✓ Collect, record and verify data.

✓ Design new machines, media and technologies.

✓ Gather and organize relevant information to formulate an argument.

Sciences

✓ Formulate a testable hypothesis and explain it using scientific reasoning.

✓ Design a method for testing a hypothesis, explaining how to manipulate the variables and how data will be collected.

✓ Organize and present data in tables ready for processing.

✓ Interpret data to draw justifiable conclusions.

✓ Use appropriate scientific conventions to visually represent abstract ideas.

GLOSSARY

Dependent variable the variable in which values are measured in the experiment.

Independent variable the variable that is selected and manipulated by the investigator in an experiment.

Valid a valid or reasonable conclusion is a conclusion supported by a valid experimental design.

COMMAND TERMS

Comment give a judgment based on a given statement or result of a calculation.

Construct display information in a diagrammatic or logical form.

Outline give a brief account or summary.

✓ Organize and present information using appropriate scientific terminology.

✓ Evaluate the validity of the method.

✓ Describe improvements to a method, to reduce sources of error.

| OTHER RELATED CONCEPTS | Consequences Evidence Models Patterns |

Introducing interaction

The global pharmaceutical industry generates a trillion dollars a year, placing it among the world's most profitable businesses (Table 9.1).

Company	2013 Revenue (US$)	Biggest selling drug
Johnson and Johnson	$71.3 billion	Remicade®—arthritis
Novartis	$57.9 billion	Diovan®—high blood pressure
Roche	$52.3 billion	Actemra®—arthritis
Pfizer	$51.6 billion	Lipitor®—cholesterol
Sanofi	$45 billion	Plavix®—heart disease
GlaxoSmithKline	$44 billion	Advair®—emphysema

Table 9.1 Revenue from the biggest pharmaceutical companies—2013
(www.fiercepharma.com/special-reports/top-10-pharma-companies-2013-revenue)

Look again at the conditions treated by the best-selling drugs of these companies: every one is a treatment for a condition that becomes more prevalent as populations live for longer.

Lipitor®—a treatment for high cholesterol levels—has made Pfizer US $131 billion since its release. It works by inhibiting the natural production of cholesterol within the body (Figure 9.1). Inhibition is one of the key ways that biological molecules and drugs interact with each other in order to regulate the production and uptake of important substances in the body.

Maintaining baseline levels of important chemicals in an organism's internal environment is often achieved through a series of chemical reactions that are catalysed by a series of enzymes. Enzymes are protein molecules with a particular shape, which allows smaller molecules to fit onto them.

Figure 9.1 Lipitor belongs to the family of drugs called statins, which lower blood cholesterol levels

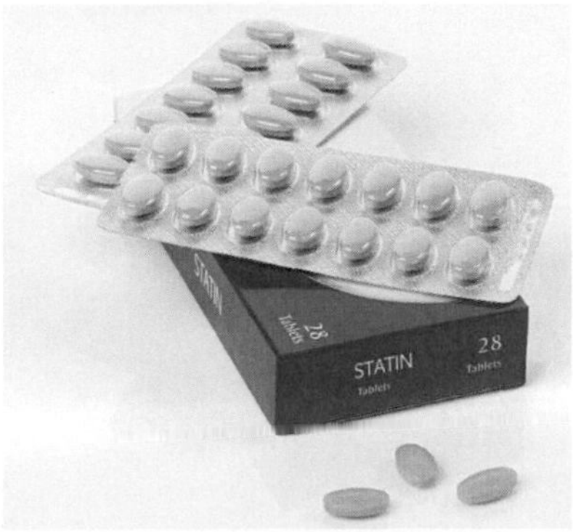

QUICK THINK

One of the big criticisms levelled at the biggest pharmaceutical companies— also known as Big Pharma—is that many of their life-saving drugs are patented, making them very expensive and unavailable for all but the most well-off or those with the best health insurance.

Very often there are generic alternative drugs available for a fraction of the cost but these are illegal to sell until the company's patent expires— which could be up to 20 years or more in some cases.

To what extent should we consider economic and intellectual property rights when human lives could be at stake?

⌘ INTERDISCIPLINARY LINKS
Individuals and Society

You may also consider issues of choice, equity, ethics and global interactions in an economics inquiry.

In these enzyme-catalysed reactions the product of one enzyme is often the reactant of the next and so on until the final stage in the pathway. The final product interacts with the first enzyme of the pathway. Often the product binds to the first enzyme and causes it to change shape thus inhibiting (blocking) the first enzyme from working. The process is called negative feedback or feedback inhibition (Figure 9.2).

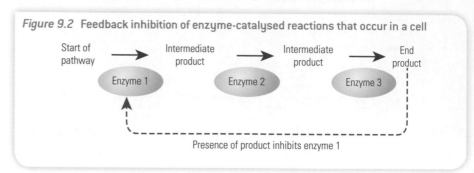

Figure 9.2 Feedback inhibition of enzyme-catalysed reactions that occur in a cell

In this way, the product of a series of reactions becomes the inhibitor that shuts down its own production if levels start to rise. And, when the product starts to be in short supply in the body, it is released from the first enzyme and lets the pathway start up again.

Knowledge of the ways in which reactants, products and enzymes interact is essential for drug companies looking to design new medicines. In the 1960s and 1970s, biochemists searched for a substance that would inhibit just one of the 30 enzymes involved in the production of cholesterol in the body. Inhibitors of this key enzyme were termed statins. The first commercial statin was approved in 1987.

Science has many examples of interactions. In science, interaction is defined as the effects that systems, objects, substances or organisms have on one another, so that the overall result is not simply the sum of the separate effects. The result of the interaction is quite often something new or something more that could not have existed without the interaction. You will explore this as you engage with the different topics in this chapter.

TOPIC 1

Interactions between organisms

All the different species in a location, which biologists call a community, interact with one another and with the environment. Within an ecosystem, these interactions are referred to as biotic variables. There are a number of ways one species can interact with another including predation, competition for resources and parasitism. Within-species interaction could include competition or cooperation.

When many individual organisms interact, new patterns or properties are often observed as a consequence of the interactions. Properties that are properties of the collective rather than properties of individuals are called emergent properties. The coordinated movement of a flock of birds or a school of fish is an example of an emergent property.

 CHAPTER LINKS
Chapter 2 on systems has information on other ways components of a system can interact with each other.

Investigating interactions

An important part of the inquiry process in science is the initial "wandering and wondering" phase where engaging deeply with the natural world triggers questions such as "is there a pattern to what I am observing?" or, if a pattern is obvious, "what is a possible cause of the pattern?".

Activity 1 — Forming a hypothesis about the variables affecting herbivore damage

This leaf has been predated by the larva of a leaf miner: an insect that consumes the nutritious tissue between the upper and lower surfaces of the leaf.

Walking through a forest should allow you to observe a number of leaves that have been predated by herbivorous insects. Take some time to "wander and wonder" and think of some testable questions. Design an experiment to address one of the following inquiry questions, or modify the question according to the types of plant where you live.

- Does the herbivore prefer larger leaves?
- What percentages of leaves are predated?
- Are leaves at a particular position more likely to be predated than leaves on other parts of the plant?
- Within a population, are trees near the edge of a forest or trees that line trails more likely to be predated than those further from the edge?

Here are some points to consider in your experimental design.

- Your hypothesis has to use scientific knowledge to explain how changes in the **independent variable** affect the **dependent variable** (the variable you would measure).
- How will you measure the independent and dependent variables?
- What should you do to ensure your conclusion is **valid**?
- How much data should you collect before you can say you have a sufficient amount of data?

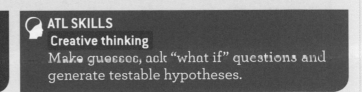 **CHAPTER LINKS**
Chapter 6 on evidence describes the features of a useful hypothesis.

GLOBAL CONTEXTS
Scientific and technical innovation

ATL SKILLS
Creative thinking
Make guesses, ask "what if" questions and generate testable hypotheses.

WEB LINKS
There are many predator-prey simulations available on the Internet. One example is found at www. nortonbooks.com/college/ biology/animations/ch34a03.htm.

This predator–prey simulation allows you to manipulate variables to see the effect on the overall predator and prey populations. Hypotheses can be created before the simulation is run. You can manipulate:

- the reproductive rate of the prey
- the number of prey organisms eaten by each predator
- the reproductive rate of the predator.

Competition can be between members of different species. This is known as interspecific competition. Some species have mechanisms for reducing this competition. For example, fungi and bacteria both compete for organic matter produced by decomposing organisms as an energy and nutrient source. But fungi produce antibiotic chemicals to kill the bacteria.

Black walnut trees and some cone-bearing trees, such as spruce, pine or cypress trees, are known to produce a toxic substance that inhibits the growth of other plants. The needles, bark and cones of some conifers have all been shown to have a negative impact on the growth of other species.

Activity 2   Investigating the effect of conifers on other plants' growth

You are going to design an experiment to investigate the effect of conifers on germinating seeds.
If the needles of a pine or spruce tree are ground with water using a pestle and mortar and the extract is filtered through cheesecloth, the resulting solution can be used to water germinating seeds.

[SAFETY] Wash your hands after contact with the extract.

STEP 1   Choose one variable to investigate. For example, needles from this season could be compared with needles from last season.

Explain why you think this variable will affect the germination of the seeds.

State the dependent variable (the variable you will measure). For example, you could measure the effect on the percentage germination, shoot growth or root growth.

STEP 2   Describe a detailed method for how you will test this hypothesis. List the materials and equipment you will need to test your hypothesis. Describe how you will alter the variables and collect data.

STEP 3   **Construct** a data table to record the data that you will collect. Include the units for the measurements.

STEP 4   Write down the procedure your group agrees on, and have it approved by your teacher.

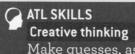

GLOBAL CONTEXTS
Scientific and technical innovation

ATL SKILLS
Creative thinking
Make guesses, ask "what if" questions and generate testable hypotheses.

Biofilms

When certain species of bacteria reach a critical concentration, individual bacterial cells coordinate their behaviour and release material that creates a matrix to which they cling as a group. The result is called a biofilm. Some biofilms are grown deliberately (eg the rind on Camembert cheese is the result of biofilm formation).

Dental plaque, the soft coating that develops on teeth, is another example of a biofilm. It illustrates two types of interaction: cooperation and parasitism. Cooperation occurs between the millions of bacteria in the mouth. Parasitism occurs because the bacteria absorb nutrients from food left on the teeth but create an acid environment that dissolves the tooth enamel, thereby harming the host.

 Activity 3   **Strategies for removing biofilms**

This child at the dentist has chewed a disclosing tablet. Disclosing tablets contain a vegetable dye that selectively dyes dental plaque. The tablets are distributed by dentists and can also be obtained from chemists (pharmacies). Using these tablets, you could calculate the percentage area covered in plaque on an individual's teeth, and investigate what factors favour the formation or removal of plaque from teeth.

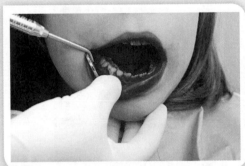

In this activity, you are going to investigate factors favouring the removal of plaque, modelling plaque on teeth by a layer of soft cheese on a rough surface.

STEP 1 Obtain the following materials:

- soft cheese
- a rough surface like an unglazed pottery plate or the reverse side of a glazed tile
- plaque disclosing tablet
- small beaker or cup
- a toothbrush
- digital camera.

[SAFETY] Do not taste or eat the cheese. Do not put the disclosing tablet in your mouth.

STEP 2 Smear cheese in a recognizable pattern on a rough surface, for example write a large capital letter with the cheese. The pattern should be just visible. Take a photograph to record the amount of cheese originally present.

STEP 3 Find various methods of attempting to remove the letter.

STEP 4 For each method, test the success of the removal of the letter using a solution of crushed disclosing tablet dissolved in water. Leave the solution on the letter for 30 seconds.

STEP 5 Take a photograph to record the amount of cheese remaining, as revealed by the disclosing tablet. Calculate the percentage of the original letter remaining.

STEP 6 Describe how effective each method of cheese removal is. Remember, this activity models the removal of plaque biofilm from teeth. The cheese is part of the model, not a real biofilm.

GLOBAL CONTEXTS
Scientific and technical innovation

ATL SKILLS
Information literacy
Collect and analyse data to identify solutions and make informed decisions.

TAKE ACTION

Organizations like the Nature Conservancy in the United States purchase land to set aside as nature reserves. Governments often establish national parks for the same purpose. The act of preserving habitat maintains individual species and also offers the opportunity for maintaining species interactions. Tree planting, removal of invasive species, clerical work, fundraising, petitioning or campaigning are all ways in which you can get involved in supporting habitat set-aside programmes.

WEB LINKS

To explore some modern materials that have been made as a result of increased knowledge of chemical bonding and materials science search for "a quick guide to smart & modern materials" on www.youtube. com.

Reflection on Topic 1

In this section, we focused on the concept of interaction between organisms. Interactions can be supportive, neutral or harmful to the individuals concerned. When you carry out scientific investigations in the future, reflect on the complexity of the interactions as you try to identify cause and effect relationships.

- How did the complexity of these interactions affect the investigations you designed in this section?

TOPIC 2

Chemical bonding

Materials can be solids, liquids or gases, plastics or metals, and can vary in strength. What role does chemistry play in material science? How would our lives be different if the interactions that allow compounds to form were not there?

There are electrostatic forces of attraction between the atoms or ions of a substance, and even between uncharged molecules. It is this interaction between atoms, ions or molecules that creates a chemical bond. Some of these interactions are stronger than others, explaining why some types of chemical bond are much stronger (harder to break) than others. These interactions allow for the formation of many different types of compound and material.

It is the valence electrons (electrons in the outer shell of an atom) that participate in the formation of a chemical bond. The number of valence electrons determines the number of atoms with which an atom can combine. The first electron shell can only hold two electrons, but for the first 20 elements of the periodic table the maximum number of electrons in the second and third shells is eight. The unreactive noble gases neon and argon have eight valence electrons, called an octet of electrons. Most atoms with an incomplete octet achieve stability by forming bonds with other atoms. They do this by gaining, losing or sharing valence electrons to obtain an octet.

An ionic bond usually results from the electrostatic attraction between oppositely charged ions formed when metallic ions transfer electrons to non-metallic ions.

A covalent bond is usually formed when electrons are shared between two non-metal atoms. The degree of sharing depends on electronegativity values for each atom.

In metallic bonding, positive metal ions are surrounded by a "sea" of valence electrons. These electrons are free to move around the metallic lattice (crystal shape), which accounts for the high conductivity of metals. The arrangement of the positive ions in layers accounts for the malleability (ability to be flattened) and ductility (ability to be drawn into thin wires) of metals (Figure 9.3).

CHAPTER LINKS
Electron shells are described in Chapter 8 on models and in Chapter 14 on pattern

Figure 9.3 Metallic bonding

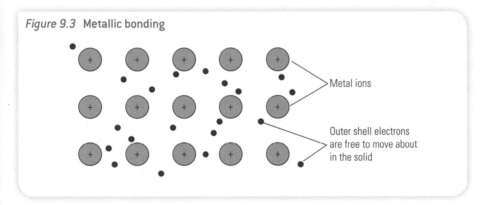

Metal ions

Outer shell electrons are free to move about in the solid

Activity 4 · Chemical bonds

You are going to make generalizations about chemical bonding through analysing pictorial representations of bonding.

STEP 1 Examine the two images below and consider the electrostatic attraction as the two atoms approach. **Outline** what is happening in each by writing a caption for each image.

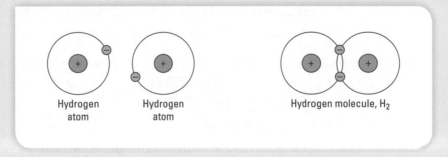

Hydrogen atom · Hydrogen atom · Hydrogen molecule, H_2

STEP 2 Predict whether H_3 would exist as a molecule.

Examine the following two images. By writing a caption for each image, explain the covalent bonding taking place between the two oxygen atoms.

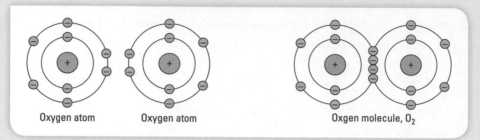

Oxygen atom Oxygen atom Oxgen molecule, O$_2$

STEP 4 Examine the image opposite. Write a caption to explain the covalent bonding that takes place between a nitrogen atom and a hydrogen atom. Use your answer to sketch a picture of the bonding in a molecule containing the elements nitrogen and hydrogen.

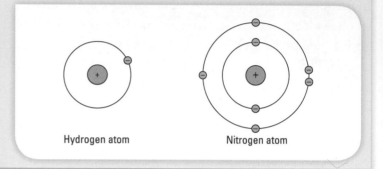

Hydrogen atom Nitrogen atom

STEP 5 Examine the following image and describe what is happening in the formation of sodium chloride, NaCl.

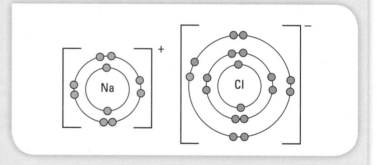

🌐 **GLOBAL CONTEXTS**
Scientific and technical innovation

🧠 **ATL SKILLS**
Critical thinking
Draw reasonable conclusions and generalizations.

In network covalent bonding (giant covalent), separate molecules are not formed. Instead a continuous network of atoms arises as a result of the interactions between many different adjacent atoms. This type of bond is the strongest bonding force (Figure 9.4).

Figure 9.4 Network covalent bonding (giant covalent) is found in diamond (carbon) and sand (silicon dioxide)

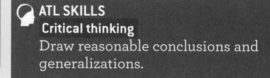

SiO$_2$

⚪ = Silicon atom
⚫ = Oxygen atom

Activity 5 Physical properties of covalent and ionic compounds

There are two main types of bonding that form as a result of two different types of interactions between electrons. Do these different types result in different chemical properties?

In this experiment, you will compare and contrast the physical properties of an ionic compound (sodium chloride—table salt) and a covalent compound (camphor). Your teacher may provide you with other samples of ionic and covalent compounds to use. You will investigate the following properties: odour/volatility, hardness, solubility, melting point and conductivity and use the results of these tests to distinguish between ionic and covalent compounds.

STEP 1 Gather together the following materials:

- eye protection
- paper towels
- watch glass
- two 250 cm³ beakers
- two test tubes
- 10 cm³ graduated cylinder
- conductivity meter
- distilled water
- Bunsen burner and ring stand apparatus or hot plate
- heatproof mat
- tongs.

[SAFETY] Wear eye protection. Wash your hands after touching the chemicals. Do not heat the substances for more than 30 seconds. Do not let the camphor ignite. If it does, stop heating, cover the beaker with a heatproof mat and alert your teacher.

STEP 2 **Construct** a data table in which you can compare the properties and characteristics of salt and camphor for the following tests.

STEP 3 Place a few crystals of sodium chloride and camphor on separate pieces of paper towel. Observe the odour of each solid by wafting the sample. Save these samples to be used for the hardness test.

STEP 4 Rub small samples of each solid between your fingers. Note whether each feels soft or hard. Check by attempting to crush a few crystals of each solid between a spoon or scoopula and a watch glass.

STEP 5 Set up a heat source (Bunsen burner and ring stand apparatus or hot plate). Place a pinch of camphor in a clean, dry 250 cm³ beaker. Gently heat this beaker over the heat source for no more than 30 seconds. Using a new beaker, which is cool, repeat for a pinch of salt and record your observations.

STEP 6 Place one finger-width (about 2 cm³) of water in two separate test tubes. Add a pinch of camphor to one and a pinch of salt to the other. Swirl or shake each test tube for about 2 minutes.

STEP 7 Using a clean, dry 250 cm³ beaker, add a pinch of camphor and then add 10 cm³ distilled (not tap) water to the beaker and stir to dissolve. Place the conductivity meter in the beaker and observe. Repeat, this time using salt.

STEP 8 Based on the results from your investigation, distinguish between ionic and covalent compounds.

🌐 GLOBAL CONTEXTS
Scientific and technical innovation

🧠 ATL SKILLS
Critical thinking
Draw reasonable conclusions and generalizations.

⛓ CHAPTER LINKS
See Chapter 8 on models for a more in-depth look at the concept of models.

The interactions between atoms—chemical bonds—cannot be seen or watched because atoms are very small. About 5 million million hydrogen atoms can fit on the top of the head of a pin.

But we know these interactions exist, because compounds such as water and salt, and materials such as silk and wood, could not exist without them. So, how do we go about representing something we cannot see? For phenomena that they cannot see, scientists usually make use of models.

Some models are simple while others are very complex, but all are based on theoretical understanding.

👤 Activity 6   Constructing molecular models

Before you begin this activity, review your answers to Activity 4.

In this activity, you will use a molecular model kit to construct models of molecules and, as a result, construct a visual representation of the interactions (bonds) for the different types of covalent bond. The use of molecular model kits will help to demonstrate the 3D shapes of molecules and the interactions between atoms in these molecules.

Use as many connectors as possible to join the atoms in each of the following molecules. Keep in mind that each connector represents a separate interaction between atoms. Sketch the models that you construct.

- H_2, Cl_2, Br_2, HCl and HBr
- methane (CH_4) and carbon tetrachloride (CCl_4)
- ammonia (NH_3)
- water (H_2O)
- oxygen (O_2), carbon dioxide (CO_2) and ethene (C_2H_4)
- nitrogen (N_2), hydrogen cyanide (HCN) and ethyne (C_2H_2)
- ethanoic acid/vinegar (CH_3COOH)
- benzene (C_6H_6).

⛓ WEB LINKS
To explore a database of pre-constructed models, search for "molecule shapes" at phet.colorado.edu.

To build your own molecules with this simulation, search for "build a molecule" at phet.colorado.edu.

To examine the 3D structure of a variety of common chemicals, search for "structures and 3D molecules database" at www.3dchem.com/atoz.asp.

a) Why do you think you were instructed to use as many connectors as possible to join atoms in each model?

b) Describe the shapes that you observed.

GLOBAL CONTEXTS
Scientific and technical innovation

ATL SKILLS
Information literacy
Make connections between various sources of information.

You have been looking at a model of chemical bonding as the interactions between atoms, which are actually the electrostatic interactions between electrons. Are these different interactions always equivalent?

The idea of electronegativity can help us examine the equivalency of these interactions. Electronegativity is the measure of the relative attraction that an atom has for the shared electrons in a covalent bond. The electrons shared between atoms in a covalent bond are not necessarily equally distributed.

Any difference in electronegativities in a bond results in the formation of a polar bond. This means that one end of the bond is slightly positive and the other slightly negative.

Thus, the interactions between atoms do not always allow for classification of bonds into one of the two categories, ionic or covalent. Bonds cannot usually be labelled as 100 per cent covalent or 100 per cent ionic. Instead, chemical bonds are said to exist on a bonding continuum between ionic and covalent (Figure 9.5).

Figure 9.5 Bonding continuum

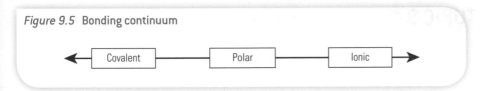

The bigger the difference in electronegativities the more ionic character a chemical bond has. The smaller the difference in electronegativities the more covalent character a chemical bond has.

 **Activity 7**  **Bonding continuum**

Through the use of examples, explain how the bonding continuum can be used to show that interactions in a chemical bond are very rarely 100 per cent covalent or 100 per cent ionic.

Comment on the statement that "scientific concepts do not cover all possibilities and should not be considered truth".

Can you come up with another way to classify the types of interaction that are present in a chemical bond? Suggest an alternative to the bonding continuum given in Figure 9.5 above.

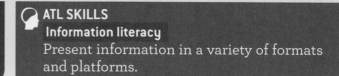

🌐 **GLOBAL CONTEXTS**
Scientific and technical innovation

🧠 **ATL SKILLS**
Information literacy
Present information in a variety of formats and platforms.

Reflection on Topic 2

Without interactions between electrons, bonds (and therefore compounds) would not exist. Modern chemists can manipulate these interactions to design new materials and new drug molecules. They do this by using computer technology to manipulate and trial different types of interaction and then try to produce the compounds in the laboratory.

- Think about what your life would be like if interactions between atoms and in molecules did not exist.

- How sure can we be of our current understanding of interactions between atoms when they occur at the nanoscopic level (at distances of 10^{-9} m)?

TOPIC 3

Electromagnetism

From small magnets in headphones, to magnets used to levitate high-speed trains, industrial magnets are everywhere. In this topic you will learn about the interaction between electric currents and magnetic fields, and how this relationship is harnessed to build large electromagnets that can be turned on and off. It is with these electromagnets that electricity is generated to power our homes, cities and factories.

Magnets

Magnets attract some metals (eg cobalt, iron and nickel). We say these metals become magnetized. The atoms in such metals are arranged in small units called domains. These domains act like smaller magnets, lining up to produce a north pole and south pole, thus becoming attracted or repelled by nearby magnets.

A magnet creates around itself a magnetic field. Objects coming near the field can become magnetized if most of the domains align with the magnet (Figure 9.6). Not all materials have domains and can act this way. This is why only certain metals can be magnetized. In addition, some materials may be magnetized temporarily in a strong magnetic field, but once the field is removed, the domains are no longer aligned and the material loses its magnetized property.

Figure 9.6 Magnetic domains inside iron before and after magnetization

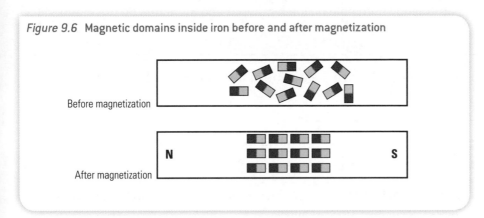

Before magnetization

After magnetization

You will discover the effects of a magnetic field in Activity 8. You will notice that opposite poles attract and like poles repel. This is due to the alignment of their magnetic fields.

Activity 8   Exploring magnetic interactions

Part A: Magnet play

You are going to explore the effect of magnets on a compass needle, and investigate the magnetic field lines created by a single bar magnet and by a pair of bar magnets.

STEP 1   Locate the following materials:

- 2 bar magnets
- compass
- one salt shaker of iron filings
- 3 pieces of paper
- pencil
- funnel.

Do not put magnets close to computers and other electronics as they may interfere with magnets inside these devices. Magnets should also be kept away from any credit or bank cards as the magnetic strip on these cards can be damaged.

TIP

Bringing a compass needle close to magnets may cause the needle to reverse polarity. To realign the compass needle domains, place the compass on a horizontal surface. Move the end of the bar magnet along the needle. Move the magnet away from the compass, and it should function properly again.

STEP 2 A compass needle points to the magnetic North pole. Bring a compass close to the north end and the south end of the magnet. What do you notice?

[SAFETY] Avoid skin contact with the iron filings as they can irritate. Also, do not blow the filings as off the paper near anyone's eyes.

TIP

It is very difficult to clean a magnet covered with iron filings. Ensure there is always a piece of paper between the magnet and iron filings, or wrap the magnet in clear plastic wrap.

STEP 3 Place a compass beside each bar magnet. The point of the compass will point towards the north pole of the magnet. Note the north and south pole on each magnet and mark this on the magnet if it is not already marked.

STEP 4 Place a sheet of paper over the bar magnet. Trace the outline of the magnet on the sheet, recording the north and south poles.

STEP 5 Using the salt shaker, gently sprinkle the iron filings on top of the paper. Lightly tap the sheet of paper to help move the iron filings uniformly over the magnetic field.

Carefully sketch the shape of the magnetic field lines on the sheet. Make a note of the shape of the lines formed by the iron filings, as well as their spacing.

STEP 6 When your sketch is complete, lift the paper carefully to avoid spilling any iron filings and transfer the filings back into the salt shaker. You may need to use a funnel.

STEP 7 Place the paper back over the magnet, and move the compass over a few of the field lines you have drawn. What do you notice about the direction of the compass point? What force is acting on the compass point?

STEP 8 Repeat steps 3–6 with two magnets set up in the configurations in the diagram below: north to south and south to south. What do you notice about the direction of the field lines in each case?

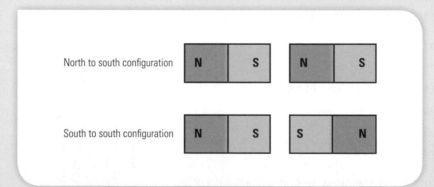

Questions

a) What type of force did you feel from the magnets in each configuration?
b) What direction do you think these field lines "flow" in?

c) What do these field lines tell you about the interaction between two magnets?

d) What do field lines represent?

e) What does the distance between the field lines tell you about the force in a magnetic field?

GLOBAL CONTEXTS
Scientific and technical innovation

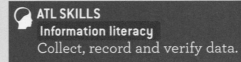

ATL SKILLS
Information literacy
Collect, record and verify data.

You may be wondering, "Can a magnet be turned off?" Magnets that can be turned on and off are called electromagnets. To understand how they work, you need to know that an electric current creates a magnetic field, due to the changing electrical field around a wire as electrons move through it. If the electrons stop moving, the magnetic field disappears because the electric field is no longer changing. Hence, electromagnets are magnetized only when there is a current. The magnetic field created by an electromagnet can be weak or strong, depending on several factors that you will explore in Activity 9.

Electric generators, MagLev trains and headphones all rely on the interaction between electric currents and magnetic fields.

By using a series of electromagnets along the length of a train and track, a magnetic levitation (MagLev) train can be lifted above the track, making for a very smooth and quiet ride. Electromagnets are also used to drive the train forwards. In Japan, a Maglev train that can travel up to 500 km/h is currently being developed and is planned to join Japan's fleet of high-speed trains in 2027. Once the train is levitated, electromagnets along the side of the track create alternating magnetic fields by continuously reversing the direction of the current flow. When coordinated, the alternating north and south poles push and pull the train down the track (Figure 9.7). MagLev trains can travel so fast because there is no contact, and therefore no friction, between the train and track.

Figure 9.7 **Maglev train propulsion system**

You are going to create an electromagnet and discover the properties affecting the strength of this magnet.

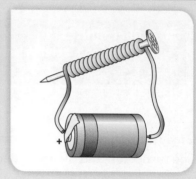

STEP 1 Collect these materials:

- 9V battery
- 0.5m copper wire (PVC coated)
- alligator clips
- iron nail
- paper clips of various sizes, loose staples
- wire stripper
- funnel.

[SAFETY] The coil of wire will get hot enough to burn skin if left connected.

STEP 2 Leaving 8 cm of copper wire at either end, wrap the wire around the iron nail in a tight coil.

STEP 3 Use wire stripper to remove the plastic coating on the ends of the wire.

STEP 4 Using alligator clips, connect one end of the wire to the battery.

STEP 5 Scatter a few paper clips and staples on the table, ready for the electromagnet to pick up.

STEP 6 When you are ready, connect the other end of the wire to the other terminal of the battery to complete the circuit. As the electrons pass through the wire, the iron nail will become magnetized. Describe what happens if you hold a nail or paper clip near the coil. Record the number of paper clips and/or staples you were able to pick up.

TIP

Leaving the wires connected will drain the battery very quickly. Therefore, do not complete the circuit unless you are ready to pick up paper clips and staples.

STEP 7 Disconnect the battery.

STEP 8 Observe how varying the design of the electromagnet affects its strength.

STEP 9 Draw the battery, wire coil and magnetic field. Label the positive and negative ends of the battery, and the poles of the coil's magnetic field.

STEP 10 How did you test the strength of your electromagnet?

STEP 11 Describe the configuration that created the strongest electromagnet.

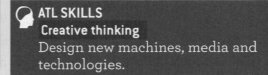

GLOBAL CONTEXTS
Scientific and technical innovation

ATL SKILLS
Creative thinking
Design new machines, media and technologies.

 Activity 10   **Electromagnet inquiry**

Now that you know how to create an electromagnet and have observed that some configurations produce a stronger electromagnet, you will design an experiment to investigate one factor affecting the strength of an electromagnet.

There are likely to be many aspects of the experiment that will need to be controlled. Brainstorm this with a partner before creating your inquiry question and procedure.

STEP 1 Choose one variable to investigate. Explain why you think this variable will affect the strength of the electromagnet.

STEP 2 Describe a detailed method for how you will test this hypothesis. Draw a diagram of your equipment setup. Describe how you will alter the variables and collect data.

State the:

- independent variable (the variable you will change)
- dependent variable (what you will measure)
- control variables (all other variables that must stay the same during trials)
- materials (list the materials and equipment you will need to test your hypothesis)
- detailed method (so that anyone can repeat your experiment).

STEP 3 Have your procedure approved by your teacher, make any changes that are needed, then carry it out.

STEP 4 **Construct** a data table to record the data that you will collect. Include the units for the measurements.

Plot your data on an appropriate graph so that the relationship between the variable you investigated and the strength of the electromagnet is easily shown.

Evaluation and conclusions

a) What can you conclude from your results?

b) How does your graph support the conclusion?

c) Does your conclusion agree with your hypothesis?

d) Evaluate the procedure and the results. Were any repeated results the same? Was your procedure **valid**? What went wrong in your plan, and why? Were there any anomalies in your results?

e) Describe what improvements to your plan you would make if you were to carry it out again.

f) Describe another experiment you could do to learn more about the problem or your prediction.

🌐 **GLOBAL CONTEXTS**
Scientific and technical innovation

🧠 **ATL SKILLS**
Critical thinking
Gather and organize relevant information to formulate an argument.

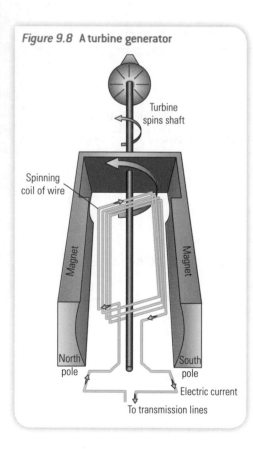

Figure 9.8 A turbine generator

Turbine spins shaft

Spinning coil of wire

Magnet

Magnet

North pole

South pole

Electric current

To transmission lines

Generating electricity

A magnetic field is produced by a moving electrical charge (a current). The reverse is also true: an electric current can be induced in wires in a magnetic field by moving the wires. In order to create a large current, there needs to be a large number of coils of wire and a way to move these wires quickly in the presence of a strong magnet.

The components of an electrical generator are shown in Figure 9.8. A turbine is used to turn the generator. The turbine can be a wind turbine or a steam turbine. The steam used by a steam turbine might be generated from heating water by burning fossil fuels or from the heat produced by nuclear fission. The turbine shaft rotates within a strong magnetic field. In this way energy is transformed from kinetic energy of the moving turbine shaft to electrical energy as an electric current is induced in the rotating coils of wire.

Reflection on Topic 3

You have learned about the effect of magnets on nearby objects, and how the interaction between electricity and magnetic fields can be harnessed to create useful technology.

- How can electricity be used to create a magnet?

- How does an electromagnet differ from a permanent magnet?

- How could a light bulb that is close to, but not touching, an electromagnet be lit?

Summary

You have learned about interactions over short and longer timescales, and on a scale from the atom to a massive MegLev train levitating above tracks and moving at hundreds of kilometres an hour. You have learned that two or more atoms, objects, organisms, substances or systems interact with one another or with the environment by transferring matter and energy. In the case of electromagnetic induction, energy can be transferred wirelessly from one circuit to another.

You have seen that the overall results of interactions are almost never a simple sum of effects but instead a complex structure.

CHAPTER 10

Consequences

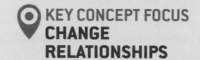

KEY CONCEPT FOCUS
CHANGE
RELATIONSHIPS

INQUIRY QUESTIONS

TOPIC 1 Environmental chemistry

- **How do changes in atmospheric chemistry affect the environment?**
- **How do our actions and choices affect the environment?**

TOPIC 2 Effect of increased greenhouse gases on global temperature

- **What is the effect of greenhouse gases on the Earth's temperature?**
- **How do increased greenhouse gas emissions cause global changes in climate?**

SKILLS

ATL

✓ Process data and report results.

✓ Make guesses, ask "what if" questions and generate testable hypotheses.

✓ Evaluate and select information sources and digital tools based on their appropriateness to specific tasks.

✓ Interpret data.

✓ Practise observing carefully in order to recognize problems.

Sciences

✓ Formulate a testable hypothesis using scientific reasoning.

✓ Design a method for testing a hypothesis, explaining how to manipulate variables, and ensure that enough data is collected.

✓ Organize and present data in tables ready for processing.

✓ Plot scatter graphs with a line of best fit and identify trends.

✓ Interpret data and explain results using scientific reasoning.

✓ Distinguish between correlation and cause and effect.

✓ Critically evaluate information from various sources, demonstrating awareness of limitations, misrepresentation or lack of balance.

✓ Describe possible extensions to the method for further inquiry.

✓ Use appropriate scientific terminology to make the meaning of your findings clear.

OTHER RELATED CONCEPTS

Balance   Environment   Interaction

GLOSSARY

Control or control group a version of the experiment in which the independent variable being tested is not applied and cannot influence the results.

Extensions to the method developments for further inquiry as related to the outcome of the investigation.

Prediction an expected result of an upcoming action or event.

COMMAND TERMS

Formulate express precisely and systematically the relevant concept(s) or argument(s).

Summarize abstract a general theme or major point(s).

Introducing consequences

With any new scientific innovation comes a promise of progress, of improvement and of benefits. But new technologies often come with risks as well.

For every new device or technique that claims to revolutionize the way humanity lives, there is a Chernobyl, or a *Titanic*, or a Space Shuttle *Columbia* to remind us that safety is never guaranteed. The arrogance of proclaiming something is "fool-proof" can have deadly consequences.

When considering how best to use new technologies, or if they should be used at all, we have to know enough about the dangers to make a decision on whether or not the risks are low enough to accept.

In science, consequences are defined as the observable or quantifiable effects, results or outcomes correlated with an earlier event or events. But what about when the consequences are unknown or unknowable? This might be because scientists lack the ability to measure the consequences, or because the processes are not completely understood. How can we be sure our judgment is not biased by focusing on the potential benefits rather than potential dangers? And how should scientific advisers to governments make decisions when the consequences are far removed from their own lifetime?

These questions are being asked now as the worldwide debate on genetic modification in agriculture continues. One of the companies at the centre of this debate is the multinational biotechnology company Monsanto. This company patents and sells genetically modified (GM) seed for many important food crops such as maize (Figure 10.1) and soybeans. Monsanto's most successful GM seeds are called Roundup Ready®. This is because they have been modified to carry a gene that makes them less susceptible to the herbicide (weedkiller) called Roundup®, which is also manufactured by Monsanto.

Because the crops are immune to herbicide, farmers can use as much Roundup® as they like on their fields to kill weeds that compete with the crop. Killing the competing weeds increases the yield of the corn or soybean crop. Monsanto calls this "a perfect fit with the vision of sustainable agriculture and environmental protection".

Many farmers and some members of the scientific community have warned about the unintended consequences of genetically modified seeds. These include food allergies, which might develop in some people because the inserted gene carries instructions for the plant to make a new protein that could be entirely new to human diet. Other concerns include that pollen from GM crops could be transferred to wild plants nearby, leading to the spread of herbicide-resistant superweeds.

What do you think? Even if there were consequences, would the benefits outweigh them?

> *I conceive that pleasures are to be avoided if greater pains be the consequence.*
>
> Michel de Montaigne

Figure 10.1 Genetically modified maize may have unintended consequences

WEB LINKS

There are an increasing number of citizen science projects where members of the public can collect and analyse data. The Project BudBurst website is one. It can be found at budburst.fieldscope.org/map/24.

Find out about others by searching for "Galaxy Zoo", "ClickToCure" or "NestWatch".

In this chapter, you will explore the concept of consequences, examining the effects of human activity on the environment and focusing on global changes in the chemistry of our atmosphere caused by burning fossil fuels. Try to keep the idea of acceptable risk and quantifiable effects in mind.

TOPIC 1

Environmental chemistry

An environment can be defined as all the external conditions that affect an organism (plant or animal, including humans) during its lifetime. Human activity affects all parts of the environment including the air that we breathe, the water that we drink and the soil in which our food grows. Not only do humans influence the environment, but our actions also influence all other organisms living on Earth, from the tiniest algae floating in the Pacific Ocean, to ladybird beetles in the tulip fields of the Netherlands, to the polar bear found in the Canadian arctic.

Due to rapid increases in the use of technology, the human race has caused great destruction to the global environment. For example, the manufacture and release into the atmosphere of CFCs damaged the ozone layer that protects life on Earth from harmful ultraviolet radiation. Deforestation and the use of coal-burning power stations have contributed to a rise in mean global surface temperature due to the increased amount of atmospheric greenhouse gases (Topic 2). A growing global population, unsustainable resource use, and land and marine pollution will continue to have consequences.

Many scientists, including NASA's director of the Goddard Institute for Space Studies, climatologist James Hansen, consider we have reached a point where the situation has become irreversible—a so-called tipping point.

Air pollution

The atmosphere is defined as a layer of gases surrounding the planet's surface. The troposphere (the gases in the first 12 km above the Earth's surface) contains the majority of the air that we breathe and is vital to the health of living organisms on the planet.

This air contains 78 per cent nitrogen (N_2), 21 per cent oxygen (O_2), 0.9 per cent argon (Ar), 0.04 per cent carbon dioxide (CO_2) and about 1 per cent water vapour (H_2O).

Air pollution occurs when the natural composition of the atmosphere changes by the addition of other gases (primarily carbon monoxide (CO), nitrogen dioxide (NO_2), sulfur dioxide (SO_2) and particulates

INTERDISCIPLINARY LINKS
Individuals and societies

You may also consider issues of ethics or choice in economics or business management.

WEB LINKS

Can we do anything to help save the environment?

Consider this while watching this short clip created during an IB World Student Conference. Go to http://www.youtube.com/watch?v=fQhCo5WXiJQ&feature=share.

QUICK THINK

Much of the information that we are exposed to about our environment comes from secondary sources. These sources include newspapers, television, movies and the Internet.

Together with a partner, discuss the following questions:

a) What is the quality of the information from these types of source?

b) What would be characteristics of an objective, high-quality source of information?

(small pieces of solid or liquid matter). High levels of atmospheric particulates are thought to harm health due to inflammation of the lung. Particulates with a diameter of 10 μm or less are called PM10. They are too small to be filtered out by nasal hairs.

 Activity 1 | **Comparing differences in particulate air pollution**

The map below shows the annual levels of atmospheric PM10 particulates in different urban areas of the world. The data is for cities with more than 100,000 inhabitants.

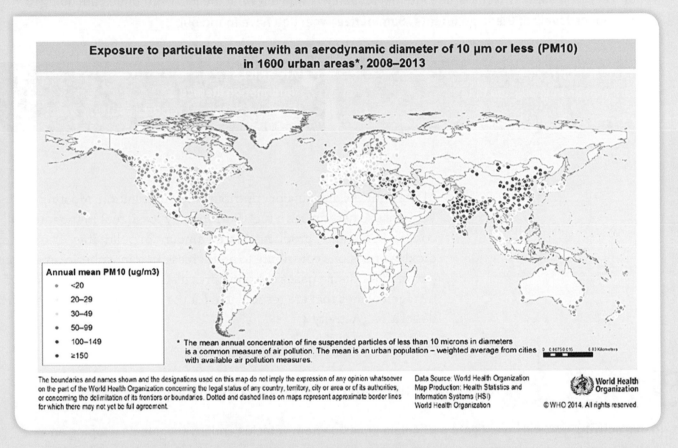

Exposure to particulate matter with an aerodynamic diameter of 10 μm or less (PM10) in 1600 urban areas*, 2008–2013

Annual mean PM10 (ug/m3)
- <20
- 20–29
- 30–49
- 50–99
- 100–149
- ≥150

* The mean annual concentration of fine suspended particles of less than 10 microns in diameters is a common measure of air pollution. The mean is an urban population – weighted average from cities with available air pollution measures.

The boundaries and names shown and the designations used on this map do not imply the expression of any opinion whatsoever on the part of the World Health Organization concerning the legal status of any country, territory, city or area or of its authorities, or concerning the delimitation of its frontiers or boundaries. Dotted and dashed lines on maps represent approximate border lines for which there may not yet be full agreement.

Data Source: World Health Organization
Map Production: Health Statistics and Information Systems (HSI)
World Health Organization

World Health Organization
©WHO 2014. All rights reserved.

When you compare the differing amounts of atmospheric particulates in developing and more developed countries, what do you notice?

 **WEB LINKS**
To take a closer look at this map visit the World Health Organization website at www.who.int and then search for "Exposure to ambient air pollution" and "maps".

 **WEB LINKS**
Access the Smog City 2 website at www.smogcity2.org and search for "Save Smog City 2 from Particle Pollution!".

Questions

a) Examine the relationship between particulates and smog formation in the simulation referred to in the web link. Use this and other research to describe the factors that could influence the different annual readings shown in the map.

b) Choose two cities, one with low levels of particulates and one with high levels of particulates, in different geographic locations on the map. Investigate hourly and daily trends for levels of particulates and ozone (O_3), NO_2, SO_2 and CO by searching the air quality index database http://aqicn.org for your chosen cities.

c) For these same two cities, use internet research to find out what the local authorities are doing to reduce levels of these pollutants. **Summarize** what you have found out.

GLOBAL CONTEXTS
Orientation in space and time

ATL SKILLS
Information literacy
Process data and report results.

Motor vehicles are a major contributor to air pollution. Most motor vehicles use gasoline as a fuel. Burning any fossil fuel in the internal combustion engine produces a large amount of pollutants. The carbon dioxide emissions contribute to an increase in atmospheric greenhouse gases. The exhaust gases contain particulates and sulfur dioxide and nitrogen oxides that are responsible for the formation of smog and acid deposition (Activity 4).

Activity 2 — Measuring particulates produced by different vehicles

In this activity, you will investigate particulate emission from vehicles and investigate developing an anti-idling policy at your school.

STEP 1 Design a complete and safe experiment to investigate the amount of particulates produced by different vehicles.

STEP 2 List the variables that you will have to consider.

STEP 3 **Formulate** a hypothesis that relates to this investigation.

STEP 4 **Summarize** the scientific, social or technological concerns or opportunities that could be addressed.

STEP 5 Suggest how you could make **extensions to the method** to make further or related inquiries.

Another pollutant formed in vehicle exhaust is carbon monoxide. The graph below shows carbon monoxide production of two cars at start-up and after 5 minutes.

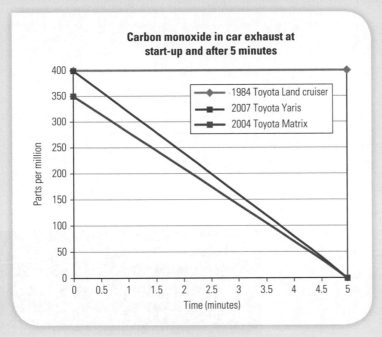

Carbon monoxide in car exhaust at start-up and after 5 minutes

Legend:
- 1984 Toyota Land cruiser
- 2007 Toyota Yaris
- 2004 Toyota Matrix

(y-axis: Parts per million, 0 to 400; x-axis: Time (minutes), 0 to 5)

TAKE ACTION
Does your school community have an anti-idling policy? If not, investigate developing one.

STEP 6 Explain the relationships that are shown in the graph.

GLOBAL CONTEXTS
Scientific and technical innovation

ATL SKILLS
Creative thinking
Make guesses, ask "what-if" questions and generate testable hypotheses.

Activity 3 — Electric vehicles—viable or not?

One possible solution to combat the problem of air pollutants created by vehicles burning fossil fuels is to use electric vehicles. Is this a viable solution? Are these vehicles better for the environment than vehicles with internal combustion engines?

STEP 1 Read and analyse three different types of Internet resource (see web links below).

STEP 2 Comment on how each source is appropriate for different purposes (eg for MYP Sciences research or for Individuals and societies research).

STEP 3 Comment on how reliable and objective these sources are.

STEP 4 Select some suitable information sources to assess whether electric vehicles are a viable solution to solving the problem of atmospheric pollutants from vehicles with internal combustion engines.

STEP 5 Discuss the scientific evidence for this argument. Document your sources.

WEB LINKS

The three different types of Internet resource could be a blog post, a commentary from an environmental organization and a news article. Try searching the following sites to locate some articles with different viewpoints:

http://shrinkthatfootprint.com and search for electric vehicles
http://content.sierraclub.org and search for electric vehicles myth
http://www.nationalgeographic.com and search for electric vehicles alone.

You could also search for "environmental impact" and "electric vehicles".

GLOBAL CONTEXTS
Scientific and technical innovation

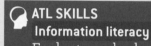

ATL SKILLS
Information literacy
Evaluate and select information sources and digital tools based on their appropriateness to specific tasks.

Acid deposition

Acid deposition refers to any form of precipitation (rain, snow, sleet, fog or hail) that may be contaminated with several different acids.

The pollutants SO_2 and NO_2 are produced by vehicle exhausts, volcanic eruptions and in fossil-fuel power stations. Carbon dioxide is present in our atmosphere as part of the carbon cycle, but the concentration is increasing due to deforestation and the combustion of fossil fuels.

These three gases react with moisture in the atmosphere to form sulfuric acid, nitric acid and carbonic acid. Precipitation is naturally slightly acidic with a pH of about 5.6 due to the presence of carbon dioxide in the atmosphere. But acid deposition is much more acidic.

QUICK THINK

A map of the US (Figure 10.2) indicates that about a third of the country has a problem with acidic precipitation.

a) Suggest what factors led to this.
b) Discuss various methods of preventing and controlling acid deposition.

Figure 10.2 Measurement of precipitation pH across the United States in 2012

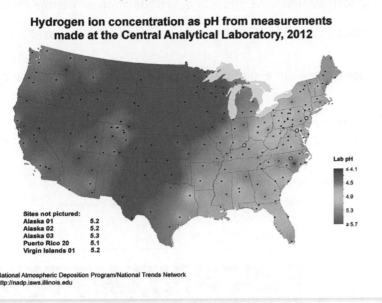

Hydrogen ion concentration as pH from measurements made at the Central Analytical Laboratory, 2012
Lab pH
≤ 4.1
4.5
4.9
5.3
≥ 5.7
Sites not pictured:
Alaska 01 5.2
Alaska 02 5.2
Alaska 03 5.3
Puerto Rico 20 5.1
Virgin Islands 01 5.2
National Atmospheric Deposition Program/National Trends Network
http://nadp.isws.illinois.edu

Activity 4 Monitoring the effects of acid deposition

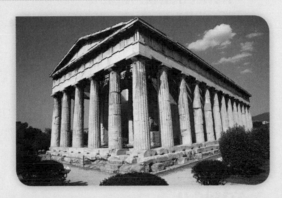

When sulfuric acid, H_2SO_4, is the acid in acid deposition, the following chemical reaction takes place between acid precipitation and a marble statue (composed of calcium carbonate, $CaCO_3$):

$$H_2SO_4(aq) + CaCO_3(s) \rightarrow CaSO_4(s) + H_2O(l) + CO_2(g)$$

Over time, the statue is damaged as the calcium carbonate is removed. The calcium sulfate formed is slightly soluble and washes away.

In this activity, you will design an investigation to collect and interpret data to determine the effect of acid deposition on marble statues.

STEP 1 With a partner, discuss the factors that would influence the amount of destruction that acid deposition has on marble statues.

Design a complete and safe method to investigate the effect of acid deposition on marble statues.

a) What are the variables that you have to consider? What variable will you measure, and what variable will you change?

b) Why is it important to include a **control** in your experiment? Ensure that you include a control in your design.

c) Formulate a hypothesis that relates to this experiment.

d) Suggest how much data you would have to collect to ensure that the method is valid and any conclusions you make are justified.

e) Construct a data table you could use to display your data so that it is easily understood.

STEP 3 Suggest how you could extend your experiment.

GLOBAL CONTEXTS
Scientific and technical innovation

ATL SKILLS
Critical thinking
Interpret data.

Reflection on Topic 1

As humans, all of our actions have a consequence on our environment, both locally and globally.

- What are some of our behaviours that we can change to reduce the negative impact?
- What responsibility do governments and global organizations have in order to ensure that the consequences of their actions are minimized?

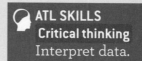

TOPIC 2

Effect of increased greenhouse gases on global temperature

Decisions that individuals, nations and governments make today about uses of technology have consequences in the future. This can easily be seen in the case of changing global climate patterns. Over the past 200 years, atmospheric carbon dioxide and other greenhouse gases have increased in concentration. This rise is due to human activities, such as burning fossil fuels and changing land use. The overwhelming majority of scientists say that this has caused the average global temperature of the Earth to rise. This is called global warming.

Global warming has consequences for our life on Earth. Global warming causes sea levels to rise. This is mainly due to water expanding as is heats up. But the increase in melting snow and ice on land results in extra water flowing into the seas.

QUICK THINK

Natural cycles also have an impact on the Earth's mean surface temperature. Research the effects of these cycles.

A rise in the sea level is a big problem for many countries, especially ones in which the land is only a few metres above sea level (low-lying land). These areas will be flooded and homes, farm land and ecosystems will be destroyed.

The climate at different locations around the world is affected by the rise in air and water temperatures. There is more evaporation and precipitation overall, which leads to droughts and desertification in certain areas of the globe, while other areas experience flooding. The impact of climate change is most striking in the arctic. In recent years, temperature increases in the arctic have been almost twice as large as anywhere else in the world.

In this topic, you will learn about the characteristics of greenhouse gases, and their effect on mean global temperature. With this knowledge, you can make informed decisions on how to decrease your own carbon footprint.

QUICK THINK
When ice that is floating in water melts, there is no change in sea level. Discuss why this is the case.

QUICK THINK
Are hurricanes getting worse because of global warming?

About greenhouse gases

The Earth's atmosphere is made of 78 per cent nitrogen, 21 per cent oxygen and 1 per cent other gases, such as argon and what are called "greenhouse gases". Greenhouse gases found naturally in the atmosphere are water vapour, carbon dioxide, methane and nitrous oxide. These gases absorb and emit infrared radiation.

Sunlight passes through the atmosphere and heats up the surface of the Earth. Some of this absorbed energy is reflected back into space as infrared radiation. Greenhouse gases in the atmosphere absorb and emit this infrared radiation and so warm the atmosphere. This, in turn, radiates energy back to the Earth. This "trapping" of the Sun's energy is called the greenhouse effect (Figure 10.3). This is a natural effect giving a constant stable temperature. Without the greenhouse gases in the atmosphere, the Earth's mean surface temperature would be around −18°C instead of the present +15°C.

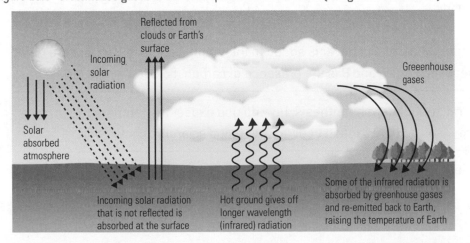

Figure 10.3 **Greenhouse gases in our atmosphere warm the Earth (the greenhouse effect)**

However, if the level of these greenhouse gases in the atmosphere increases, more infrared radiation is "trapped" in the Earth's atmosphere and emitted back to Earth. This increases the overall temperature of the Earth.

Carbon dioxide contributes to 15–20 per cent of the overall greenhouse effect. Therefore, it has a major influence on global warming. The burning of fossil fuels releases carbon dioxide into the atmosphere. Humans have increased atmospheric carbon dioxide concentration by a third since the Industrial Revolution began.

Another cause of the increase of carbon dioxide in the atmosphere is deforestation. Plants absorb carbon dioxide and release oxygen into the atmosphere. Cutting down trees leads to an increase in the level of carbon dioxide in the atmosphere.

As the global population increases, forests are being destroyed for a number of reasons: logging, to use the land for agriculture, to mine natural resources.

 Activity 5   Greenhouse effect in a beaker

In this activity, you will investigate if carbon dioxide causes an increase in temperature inside a closed beaker. You will complete the investigation using either a regular thermometer or a temperature probe.

[Safety] Wear eye protection during Part B.

STEP 1   Collect the following materials:
- 250 cm^3 beaker
- 100 cm^3 graduated cylinder
- 200 cm^3 vinegar
- weighing dish
- 2 g baking soda
- scale
- clear plastic wrap large enough to cover the top of beaker
- light box or lamp that produces a fair amount of heat (an incandescent light bulb is best)
- stopwatch (if using a thermometer)
- thermometer or temperature probe with datalogging software
- modelling clay.

Part A

Note: To make this a fair experiment, 100 cm^3 of vinegar is added to the empty beaker in Part A of the experiment, so the difference between Parts A and B is the presence of carbon dioxide.

STEP 2   Collect the materials and **formulate** a hypothesis. How do you expect the temperature in the beaker to change once CO$_2$ is added? Explain your answer using scientific knowledge.

STEP 3   Using a graduated cylinder, measure 100 cm^3 vinegar and pour into the beaker.

STEP 4 Cover the beaker with plastic wrap. Insert the thermometer/probe by lifting the wrap slightly to give enough room to slide the thermometer/probe in. Take care not to touch the metal part of the probe with your hands. One team member should hold the thermometer/probe so it does not touch the vinegar solution at the bottom of the beaker. Seal the plastic wrap around the top of the beaker and the probe. You may need a piece of modelling clay to hold the probe in place. Once the temperature reading stabilizes, record the initial temperature of the air in the beaker.

STEP 5 Plug in the lamp/light box and place the beaker 10 cm from the light source.

STEP 6 Measure and record the temperature in the beaker every 30 seconds for 5 minutes. Your teacher may instruct you to write down your own observations, or set up the datalogger to collect data automatically.

Part B

In this part of the experiment, you will mix baking soda and vinegar to produce the following reaction:

$$CH_3COOH(l) + NaHCO_3(s) \rightarrow NaCH_3COO(s) + H_2O(l) + CO_2(g)$$

vinegar   baking soda   sodium acetate   water   carbon dioxide

The carbon dioxide gas produced by the reaction will stay in the beaker because it is more dense than air. You must cover the beaker to ensure that the gas does not diffuse out during the experiment.

STEP 1 Using a graduated cylinder, measure 100 cm^3 of vinegar and pour into the beaker.

STEP 2 Using a weighing dish and scale, measure 2 g of baking soda ($NaHCO_3$).

STEP 3 Prepare a piece of clear plastic wrap that will fit over the beaker.

STEP 4 Add baking soda to the beaker and cover it quickly with the plastic wrap. The solution will start fizzing immediately.

STEP 5 Insert the thermometer/probe by lifting the plastic wrap slightly, giving only enough room to slide the thermometer/probe in. Have one team member hold the thermometer so it does not touch the vinegar solution at the bottom of the beaker. Seal the plastic wrap around the top of the beaker and the thermometer/probe. You may need a piece of modelling clay to hold the thermometer in place.

STEP 6 Measure and record the initial temperature of the contents of the beaker.

STEP 7 Plug in the lamp/light box and place the beaker 10 cm from the light source.

STEP 8 Measure and record temperature in the beaker every 30 seconds for 5 minutes.

STEP 9 Plot your findings from Parts A and B on the same graph using two lines of best fit, showing the difference between the temperature in the beaker with and without carbon dioxide.

Conclusions and evaluation

a) What can you conclude from your results? How was the temperature in the beaker affected by addition of carbon dioxide?

b) How does your graph support the conclusion?

c) Evaluate the procedure and the results. Was your experiment valid? Were there any anomalies in your results?

d) Describe what improvements to the procedure you would make if you were to carry it out again.

e) Describe another experiment you could do to learn more about the problem or your **prediction**.

f) Research some processes that produce greenhouse gases. What do you expect will occur in our atmosphere if greenhouse gas production increases?

GLOBAL CONTEXTS
Scientific and technical innovation

ATL SKILLS
Critical thinking
Practise observing carefully in order to recognize problems.

The carbon dioxide record

The longest record of direct measurements of atmospheric carbon dioxide is from Mauna Loa Observatory, Hawaii (Figure 10.4). There is usually a periodic change in atmospheric carbon dioxide between seasons.

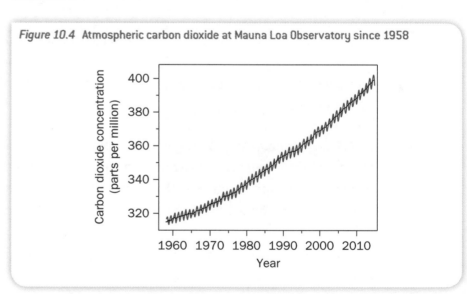

Figure 10.4 Atmospheric carbon dioxide at Mauna Loa Observatory since 1958

Plant growth reaches a maximum during the summer months, so plants take up more carbon dioxide in these months. In the winter months, less carbon dioxide is used by plants because they are not growing as much. In Hawaii, there is year-round plant growth due to the tropical climate. However, carbon dioxide spreads evenly around the world, so even on Hawaii measurements of carbon dioxide are representative of global average CO_2 concentrations.

Activity 6 Interpreting recent data on atmospheric carbon dioxide levels

Using the graph in Figure 10.4, discuss the overall trend of the concentration in carbon dioxide. Discuss what you think might be the causes and consequences of this trend.

GLOBAL CONTEXTS
Globalization and sustainability

ATL SKILLS
Critical thinking
Interpret data.

Activity 7 Interpreting the change in carbon dioxide in the atmosphere over the past 200 years

Since the start of the Industrial Age, human activities have had an effect on the atmospheric concentration of carbon dioxide and other greenhouse gases. This is shown in the two graphs below. The data lines are not straight; this shows us that the rate at which the concentration is changing is not constant, but varies.

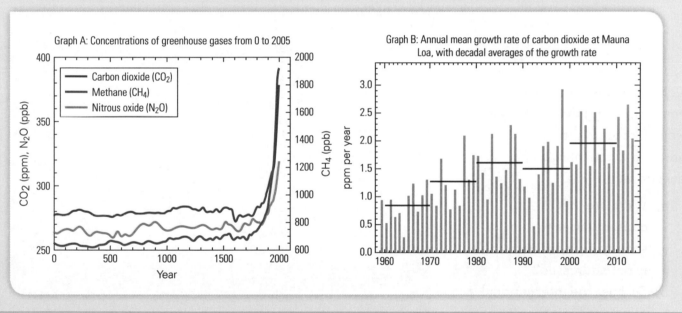

Questions

a) Find how much the concentration of carbon dioxide increased from 1800 to 1900.

b) Find how much the concentration of carbon dioxide increased from 1900 to 2000.

c) Find the percentage change in carbon dioxide concentration from 1800 to 2000.

d) Repeat questions (a)–(c) for the concentrations of methane (CH_4) and nitrous oxide (N_2O).

e) Using your answers from (a)–(d) and the two graphs, describe how the rate of change of atmospheric greenhouse gas concentration is changing.

f) From the trends seen in the graphs, what prediction could be made bout the concentration of greenhouse gases in 2030?

g) Use the simulation in the web link to discuss the relationship between the rise in carbon dioxide in the atmosphere and the surface temperature of the Earth.

WEB LINKS
Try the simulation at http://carboncycle.aos.wisc.edu/carbon-budget-tool.

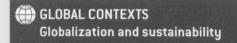

If the global output of greenhouse gases continues to grow, the Intergovernmental Panel on Climate Change (IPCC) Assessment Reports forecast some dramatic consequences. For example, in a best-case scenario, the global mean surface temperature by the end of the 21st century is likely to have increased by 0.3–4.8°C relative to the mean for 1986–2005. There will be coastal impacts as sea levels rise and public infrastructure is damaged during extreme weather events or flooding. The National Resources Defense Council in the US estimates that the average annual hurricane damage in the US alone in 2100 will be US$422 billion. In addition, as ecosystems are stressed by these changes, there will be water shortages and an increased limitation on water resources worldwide.

QUICK THINK

Conduct some research on your own using the suggested web links to investigate answers to the following questions.

a) What are the consequences to our society if global warming is not addressed?

b) Discuss the international efforts to limit the rise of greenhouse gases.

🏃 TAKE ACTION

Most people in the developed world use more energy and have higher carbon footprints than people in developing countries. What goals can you set, individually or as a school, to reduce your carbon footprint?

What can you do at home?

- Change your home's most frequently used light bulbs to energy-efficient light bulbs.
- Turn off lights you don't need.
- Don't use the air conditioner or heater unless necessary. Try putting on an extra sweater or cooling the house naturally by opening windows and doors.
- Find out if your home is well insulated.
- Reduce drafts around doors and windows.
- Walk, bike or share a ride to school, or use public transport.
- Reduce, reuse, recycle to decrease greenhouse gas production from manufacturing and disposal.
- Spread the word about how you are reducing your carbon footprint.

What can you do at school?

- Encourage recycling at your school, and hold class contests on the amount of paper recycled.
- Bring your lunch in reusable containers.
- Lobby for renewable energy to be used at your school.

Reflection on Topic 2

The majority of scientists agree that human interference with the climate system is occurring.

- How could global warming have an effect on the Earth's food supply?

- How can technology be put in place to reduce greenhouse gas emissions?

Summary

Human activities have resulted in atmospheric pollution and an increased greenhouse effect. These results have had environmental, social and economic consequences. The consequences can now be measured, and scientists are exploring ways of limiting the consequences and finding a balance between competing economic and social concerns. In the case of climate change, research includes breeding crop varieties that are more tolerant of heat and drought, as well as reducing greenhouse gas emissions by exploring wind and solar power, fuel cells, more fuel-efficient vehicles and more efficient electrical appliances.

As was mentioned in Chapter 3, the precautionary principle is one of the most important ways to think about the future consequences of new technologies and scientific innovation. For some new technologies, it may be necessary to delay implementation while the long-term consequences are explored.

You will have gained a sense of the risks of human activities and, hopefully, you will use the precautionary principle in your own life and your eventual chosen profession.

WEB LINKS
Search for an online calculator that will estimate your personal carbon footprint, which is the annual mass of carbon dioxide produced by your consumption of food, fuel used in your home or for transport, and the manufactured goods you buy and use. One such calculator that allows you to input detailed data from energy bills is www.epa. gov/climatechange/ ghgemissions/ind-calculator.html.

WEB LINKS
Visit the UN Climate Change Portal www.un.org/climatechange/ and select "Take Action".

INQUIRY QUESTIONS

TOPIC 1 Linking form to function

- **How does the density of a bone relate to its function?**
- **How can variations in the bone length of humans compared with chimpanzees be explained?**
- **How does the form of feathers vary?**
- **How does the form of leaves growing in shade differ from those exposed to light?**
- **How does the number of specialized structures on a leaf vary with position?**

TOPIC 2 Structure of organic molecules

- **Are there patterns in how organic compounds are named?**
- **How does the structural form of an organic compound relate to its physical properties?**

TOPIC 3 Forms of energy

- **Can there be a continuous transformation from one form of energy to another?**
- **How does conservation of energy link with energy transformation?**

SKILLS

ATL

✓ Access information to be informed and inform others.

✓ Make guesses, ask "what if" questions and generate testable hypotheses.

✓ Gather and organize relevant information to formulate an argument.

✓ Use and interpret a range of discipline-specific terms and symbols.

✓ Encourage others to contribute.

✓ Draw reasonable conclusions and generalizations.

✓ Use brainstorming and visual diagrams to generate new ideas and inquiries.

Sciences

✓ Formulate a testable hypothesis and explain it using scientific reasoning.

✓ Design a method for testing a hypothesis, and explain how data will be collected.

✓ Explain how to manipulate variables, and how enough data will be collected.

GLOSSARY

Parallax error a measurement error resulting from viewing the scale along different lines of sight.

Qualitative data refers to non-numerical data or information that is difficult to measure in a numerical way.

Quantitative data refers to numerical measurements of the variables associated with the investigation.

COMMAND TERMS

Formulate express precisely and systematically the relevant concept(s) or argument(s).

Justify give valid reasons or evidence to support an answer or conclusion.

Measure obtain a value for a quantity.

Plot mark the position of points on a diagram.

✓ Organize and present data in tables ready for processing.

✓ Interpret data and explain results using scientific reasoning.

✓ Use appropriate scientific conventions to visually represent molecules and name organic compounds.

✓ Plot scatter graphs to identify relationships between variables.

✓ Draw conclusions, and explain these using scientific reasoning.

✓ Describe improvements to a method, to reduce sources of error, and possible extensions to the method for further inquiry.

| OTHER RELATED CONCEPTS | **Energy   Evidence   Function   Patterns Transformation** |

Introducing form

As you have probably learned in science class, cells take on many forms. The nerve cells of the giant squid can be 12 m long, whereas the bacteria that live in our intestines are microscopically small—about 6 million times smaller. Cells are extremely varied in their appearance and their function.

In science, we define form as the features of an object that can be observed, identified, described, classified and categorized.

Biologists and medical professionals have been doing just that since Robert Hooke first used his microscope (Figure 11.1) to observe and describe the cell in the late 1600s.

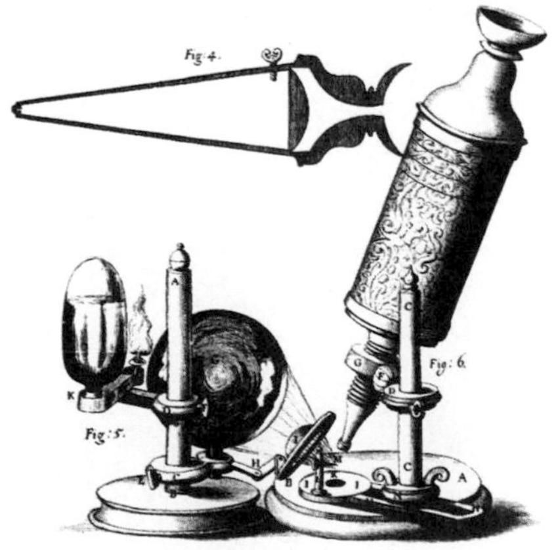

The bigger question regarding cells may not be why there are so many different forms but how these different forms are differentiated. How do they appear during development from the single original cell formed when an egg and a sperm cell join? This single cell divides into two cells that are genetically identical, each divides again and so on. So how do the 30 trillion cells in your body become so many different types of cell?

Scientists have been very interested in this question for some time. This is because, if the mechanism that tells cells to take on different forms could be discovered, we might be able to direct cells to take on the forms that we need (Figure 11.2). Imagine being able to produce a new brain or heart or muscle tissue to replace damaged or diseased tissue, or being able to treat diseases such as diabetes where a certain cell type does not function correctly.

Figure 11.1 Robert Hooke's microscope

Figure 11.2 The potential for stem cell therapies

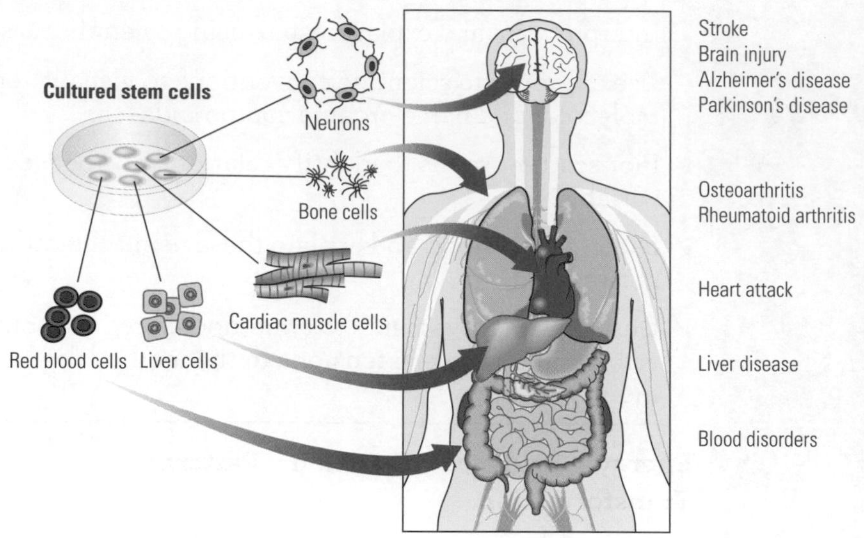

This is precisely the promise that stem cell technology provides, but, as with any new technology, there may be unknown, potentially dangerous consequences.

One of the underlying scientific questions in the stem cell debate is whether or not transplanted cells will function correctly. Having the correct form (shape and structure) is not enough if a chemical switch to turn on the necessary gene is not activated.

As you will see in this chapter, it is very difficult to separate the concepts of form and function. And, depending on the context, even seemingly small changes in form can alter function in significant and potentially negative ways.

TOPIC 1

Linking form to function

Form can refer to the shape and arrangement of parts in an organism. The form of an organism can be observed, identified, described and categorized. Form is the result of development and is therefore determined by both genetics and environment. Form in species changes as a consequence of natural selection. The forms or features that are better suited to a function in a certain environment provide a survival advantage.

There is a relationship between an organism's forms and the functions they carry out. Structural form gives clues about function; conversely, knowing what a structure's function is gives insights about its construction.

Activity 1 Comparing forearm bone structure in humans and chimpanzees

The human forelimb and the chimpanzee forelimb have similarities and differences.

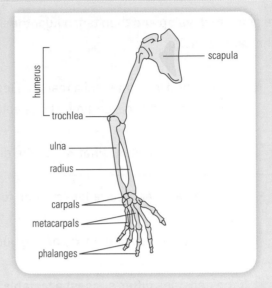

scapula

humerus

trochlea

ulna

radius

carpals

metacarpals

phalanges

STEP 1 Visit the e-skeletons interactive site at the University of Austin at Texas website www.eskeletons. org/). The website has images of bones for a variety of primates.

STEP 2 Scientists organize related organisms into groups known as taxons. Choose "taxon" and move the mouse over the chimpanzee image. Double click to select it.

STEP 3 Move the mouse over the bones of the forelimb to become familiar with the terminology.

STEP 4 Predict how the ratio of the length of bones in the chimpanzee forearm (ratio of length of humerus to length of ulna) differs from the ratio for analogous bones in humans.

STEP 5 Construct a data table to record your predictions and the measurements you will take.

STEP 6 **Measure** the lengths of the bones on screen and record the data.

STEP 7 Compare and contrast the results for chimpanzees and humans.

STEP 8 Conduct research to explain your findings. Explore some of the adaptations chimpanzees have to their environment, and suggest how your results can be explained in terms of form and function

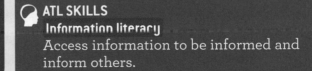

WEB LINKS
At the eLucy website www.elucy. org you can do detailed analysis on images of famous hominid skeletons.

GLOBAL CONTEXTS
Orientation in space and time

ATL SKILLS
Information literacy
Access information to be informed and inform others.

Activity 2   Predicting bone form in birds and mammals

You are going to develop and then test a hypothesis about whether analogous bones in different organisms have the same density.

STEP 1   Obtain a bird leg bone and a mammal leg bone. Collect the necessary materials according to your chosen method of determining bone volume.

STEP 2   Predict, with a reason, what will be the differences in the densities of the two bones.

STEP 3   Plan and describe a complete method for determining the density of the bones.

STEP 4   Carry out the investigation, recording your measurements in a suitable table.

STEP 5   If the densities differ, suggest a testable hypothesis to explain the differences.

🌐 **GLOBAL CONTEXTS**
Scientific and technical innovation

🧠 **ATL SKILLS**
Creative thinking
Make guesses, ask "what if" questions and generate testable hypotheses.

 **Activity 3**   Comparing the structure of different feathers

Birds have a number of different types of feather (see the diagram below). Flight feathers, counter feathers and down feathers differ markedly from each other in form and function.

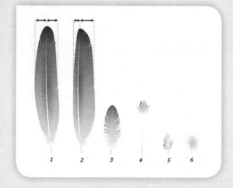

STEP 1   If feathers are available, sort them into groups using the diagram above.
Alternatively, use the web link to analyse digital images of feathers.

STEP 2   Choose three types of feather to investigate further. Collect **qualitative data** and **quantitative data** for each feather type in order to make a comparison.

STEP 3 Construct a three-circle Venn diagram to compare and contrast the properties of the feathers.

STEP 4 Conduct research into the function of each feather type and summarize how feather form is linked to function.

WEB LINKS
Go to the Arizona State University "Ask a Biologist" website at www.askabiologist.asu.edu and search for "feather zoom gallery".

GLOBAL CONTEXTS
Scientific and technical innovation

ATL SKILLS
Critical thinking
Gather and organize relevant information to formulate an argument.

Form and function in leaves

Leaves are organs of photosynthesis in plants. In order to gather the maximum amount of sunlight, leaves should have a maximum surface area. However, building leaf tissue requires energy and nutrients, so leaf form usually represents a compromise between maximizing surface area and minimizing volume.

 **Activity 4**   The surface area of shade leaves versus sun-exposed leaves

Leaves growing in shade compete with each other for sunlight. Choose one of the following hypotheses to test, and then design an investigation to test it.

- Are their mean surface areas larger in an effort to capture the same amount of light?
- Are their mean surface areas smaller in order to avoid wasting resources on creating tissues that will not capture sunlight?

Here are some points to consider.

- How many leaves must be measured to form a representative sample of the population of leaves?
- How will you decide the criteria for categorizing the leaves?
- If leaves are photographed, the area can be determined using image-processing software apps.

[SAFETY] Check with your teacher which types of leaf to collect. Avoid leaves that will harm you, such as nettles. Wash your hands when you have finished the activity.

TIP

If a clear acetate sheet containing a grid is used, you can use an erasable marker to trace the leaf onto the plastic. You can then determine the surface area by counting the grid squares. Marks on the acetate can be erased and the sheet used again.

GLOBAL CONTEXTS
Scientific and technical innovation

ATL SKILLS
Critical thinking
Gather and organize relevant information to formulate an argument.

Activity 5   Variations in the number of spines on defensive leaves

Forms that are better suited to a function in a particular environment provide a survival advantage. Plants like holly (*Ilex* sp.) and many members of the *Aloe* genus have spikes or spines on the edges of their leaves as a defence against predation by herbivores.

Does every leaf have the same number of spines on the leaf margins? If the number of spines is variable, what are the potential causes of this variation?

STEP 1   Discuss with a partner the potential causes of variation in the number of spines in these plants.

STEP 2   Using what you have discussed, **formulate** a hypothesis about the number of spines. Be sure you state a dependent variable and an independent variable, and say what the effect will be and why.

STEP 3   Design a procedure to test your hypothesis.

[SAFETY] Wear gardening gloves if you are collecting holly leaves. Wash your hands when you finish this activity.

🌐 **GLOBAL CONTEXTS**
Scientific and technical innovation

🧠 **ATL SKILLS**
Critical thinking
Make guesses, ask "what if" questions and generate testable hypotheses.

Reflection on Topic 1

In this section, we have explored the form of bones, feathers and leaves.

- In what ways does form suggest the function?
- To what extent does the environment influence form?

TOPIC 2

Structure of organic molecules

Organic chemistry used to be defined as the chemistry of living things (plants and animals), whereas inorganic chemistry was defined as the chemistry of non-living things (minerals).

Organic chemistry is now most easily described as the chemistry dealing with compounds that contain the element carbon. More than 80 per cent of all compounds contain carbon so it makes sense that there is a whole area of chemistry devoted just to the study of organic compounds. Besides the element carbon, organic compounds can contain atoms of hydrogen, oxygen, sulfur, nitrogen and the halogens.

Each carbon atom can make four bonds as it has four electrons in its valence shell. Organic chemistry is concerned with how these elements combine and what physical form the compounds take. The structure of organic compounds plays an important role in the form the molecules take and ultimately their use and function. The study of organic chemistry is complex and extremely important for many related fields of chemistry.

CHAPTER LINKS
Covalent bonding and electron shells are described in Chapter 8 on models.

Naming organic compounds

There is a huge and diverse array of organic compounds—millions and millions of them. To be able to study all these compounds, there has to be a precise method of naming them. The system of nomenclature used in organic chemistry is quite different from the system used to name other types of compound. The naming of organic compounds is based on the number of carbon atoms that are joined together in the longest continuous chain. This unique backbone of linked carbon atoms is important to the basic form of organic compounds.

The hydrocarbon family are compounds containing only carbon and hydrogen. The simplest member is methane, CH_4. Methane has only one carbon atom, which can make four bonds with four hydrogen atoms. If another carbon atom is added, the name of the hydrocarbon changes to reflect the change in the number of carbon atoms (Table 11.1).

The molecular formula for an organic compound shows the number of atoms of each element present but does not help explain how the atoms are bonded together. Instead, the formulae of organic compounds are often written as condensed structural formulae. This type of formula shows for each carbon atom the other atoms that are bonded to that carbon atom (Table 11.1).

Molecular formula	Condensed structural formula	Name
CH_4	CH_4	methane
C_2H_6	CH_3CH_3	ethane
C_3H_8	$CH_3CH_2CH_3$	propane
C_4H_{10}	$CH_3CH_2CH_2CH_3$	butane
C_5H_{12}	$CH_3CH_2CH_2CH_2CH_3$	pentane
C_6H_{14}	$CH_3CH_2CH_2CH_2CH_2CH_3$	hexane
C_7H_{16}	$CH_3CH_2CH_2CH_2CH_2CH_2CH_3$	heptane
C_8H_{18}	$CH_3CH_2CH_2CH_2CH_2CH_2CH_2CH_3$	octane

Table 11.1 The alkanes

These simple hydrocarbons all have single bonds and are called alkanes. The naming of alkanes gets a bit more complicated as branches (substituents) are added.

You are going to find the patterns in the organic chemistry naming system for different classes of organic compound. Your teacher will divide the class into groups of three.

STEP 1 Your responsibility as a group is to devise a set of nomenclature rules that would explain how the compounds in the table below are named.

Structure	Name
$CH_3-CH(CH_3)-CH_2-CH_3$	2-methylbutane
$CH_3-CH_2-CH(CH_3)-CH_2-CH_3$	3-methylpentane
$CH_3-CH_2-CH(CH_2-CH_3)-CH_3$	3-methylpentane
$CH_3-CH(CH_3)(CH_3)-CH-CH_3$	2,2-dimethylbutane
$CH_3-CH(CH_3)-CH(CH_3)-CH_3$	2,3-dimethylbutane
$CH_3-CH(CH_3)-CH(C_2H_5)-CH_2-CH_2-CH_3$	3-ethyl-2-methylhexane
$CH_3-CH_2-C(CH_3)(CH_2CH_3)(CH_2CH_3)-CH-CH_2-CH_3$	3,4-diethylhexane

Naming alkanes

STEP 2 To test the naming scheme your group has developed, name or draw the organic compounds in the table below. When your group has agreed on the answers and completed the table, ask your teacher to check your answers.

Structure	Name
$CH_3-CH(CH_3)-CH_2-CH_2-CH_3$	
$CH_3-CH(CH_3)-CH(CH_2-CH_3)-CH_3$	

Structure	Name
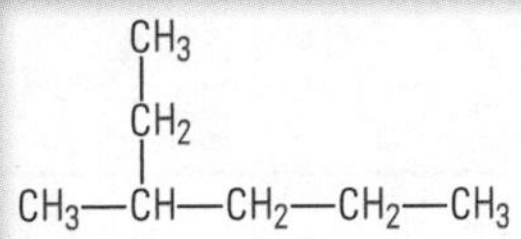 CH₃ CH₂ CH₃—CH—CH₂—CH₂—CH₃	
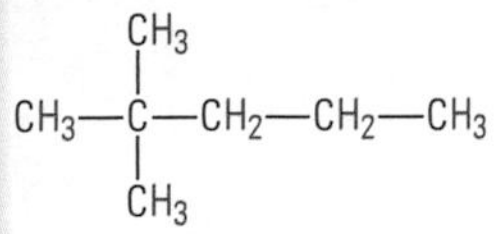 CH₃ CH₃—C—CH₂—CH₂—CH₃ CH₃	
	2,3-dimethylpentane
	3,3-dimethylpentane
	2-methylhexane
	3-ethylpentane

STEP 3 Now, number your group members 1–3. All the number 1s now join together, all the number 2s join together and all the number 3s join together.

Each new group will be assigned a different type of organic compound to develop naming rules for. Each of these different types of organic compound has a different structural unit, called a functional group, which gives that group of compounds similar chemical properties.

Once your group is confident that you have developed the naming rules for your type of organic compound, you will go back to your original groups and teach the naming rules to the other members of your group.

A Alkenes

The form of alkenes is slightly different from alkanes. Alkenes have a double bond (share two electron pairs, ie four electrons) between two of the carbon atoms in the basic backbone form.

In the following structures, only the carbon skeleton is shown for simplicity.

Structure	Name
C═C—C—C—C	1-pentene
C—C═C—C—C	2-pentene
C—C—C═C—C	2-pentene
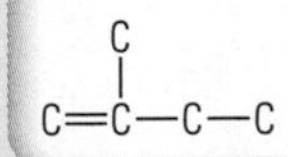 C C═C—C—C	2-methyl-1-pentene

Structure	Name
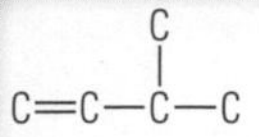 C=C—C—C with C branch on third carbon	3-methyl-1-pentene
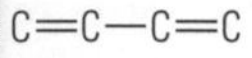 C=C—C=C	1,3-butadiene
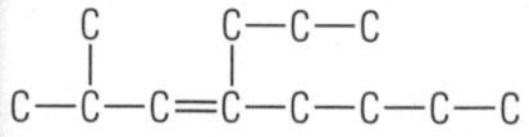 C—C—C=C—C—C—C—C with C branch and C—C—C branch	2-methyl-4-propyl-3-octene

Naming alkenes

B Alcohols

The form of alcohols is slightly different than from that of alkanes. Alcohols have a hydroxyl group (−OH) on one of the carbon atoms in the basic backbone form.

In the following structures, only the carbon skeleton and the function group are shown for simplicity.

Structure	Name
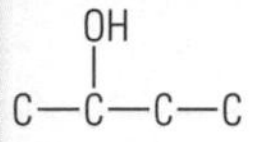 C—C—C—C with OH on second carbon	2-butanol
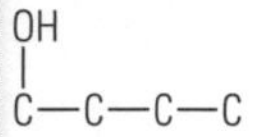 C—C—C—C with OH on first carbon	pentan-1-ol
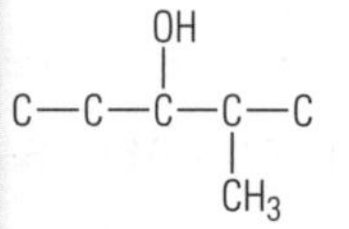 C—C—C—C—C with OH on third carbon and CH₃ below	2-methyl-3-pentanol
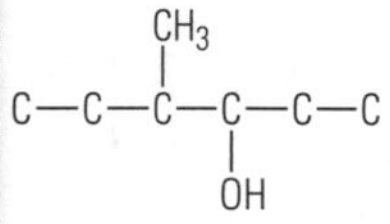 C—C—C—C—C—C with CH₃ above and OH below	4-methyl-3-hexanol
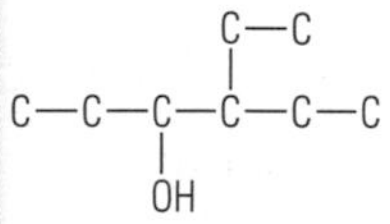 C—C—C—C—C—C with C—C branch and OH below	4-ethyl-3-hexanol
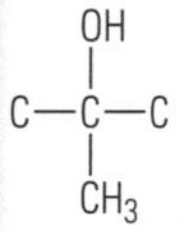 C—C—C with OH on second carbon and CH₃ below	2-methylpropan-2-ol
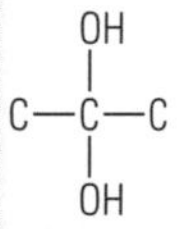 C—C—C with OH above and OH below on central carbon	2,2-propandiol

Naming alcohols

C Carboxylic acids

The form of carboxylic acids is slightly different from alkanes. Carboxylic acids have a carboxyl group, —C(=O)OH, on one of the carbon atoms in the basic backbone form.

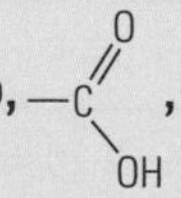

In the following structures, only the carbon skeleton and the functional group are shown for simplicity.

Structure	Name
C—C—C(‖O)—OH	propanoic acid
HO—C(‖O)—C—C—C	butanoic acid
C—C(—C)—C(‖O)—OH	2-methylpropanoic acid
C—C—C(—C—C)—C(‖O)—OH	3-methylpentanoic acid
HO—C(‖O)—C—C—C—C—C(‖O)—OH	hexanedioic acid
C—C(—C)—C(—C)—C(‖O)—OH	3,4-dimethylpentanoic acid

Naming carboxylic acids.

STEP 4 To test your knowledge of naming different organic compounds, individually complete the following table. Only the carbon skeleton and the functional group are shown for simplicity.

Structure	Name
C—C—C(—C—C top)(—C bottom)—C—C—C with C and C below	
C—C=C(—C)—C—C	
C—C—C(—C)—C—C—C(‖O)—OH	
C—C(—OH)(—C below)—C—C	
	2-methyl-1-pentene
	2,3-dimethyloctane
	2-propanol

	2-methylbutanoic acid
	2-ethyl-3-methyl-1-butanol

Naming practice of different forms of organic compounds

GLOBAL CONTEXTS
Scientific and technical innovation

ATL SKILLS
Communication
Use and interpret a range of discipline-specific terms and symbols.

Activity 7   Investigating a homologous series

A homologous series is a series of compounds that contain the same functional group (giving them similar chemical properties) but whose structures differ by a common repeating unit. Propane ($CH_3CH_2CH_3$) and butane ($CH_3CH_2CH_2CH_3$) are members of a homologous series called the alkanes. But 1-butene ($CH_2=CHCH_2CH_3$) and 1-propene ($CH_2=CHCH_3$) are members of a homologous series called the alkenes. For the alkanes, alkenes, alcohols and carboxylic acids series, members of the series each differ by a $-CH_2-$ group.

In this activity, you will explore how the physical properties of a class of compounds such as the alkanes vary when there is a slight change in form (addition of a $-CH_2-$group) across the series. You will make use of information from a database and graphical analysis to comment on any trends that you see.

Your teacher will divide your class into groups of four. Each member of your group will choose a different class of compounds (alkanes, alkenes, alcohols or carboxylic acids) to explore the changes in physical properties that occur when the form is slightly modified.

Each of you will use the online database Chemspider to find values for melting point, boiling point and solubility of the first five members (ie compounds containing 1 carbon, 2 carbons, 3 carbons, 4 carbons and 5 carbons) of your chosen class of compounds.

WEB LINKS
Chemspider is an online chemical structure database that provides text and structure search access to over 30 million structures from hundreds of different data sources. The webpage can be found at www.chemspider.com.

STEP 1 Construct a data table to record the data for your series as well as the data from your other group members.

STEP 2 By using graphing software, **plot** the data appropriately.

STEP 3 Comment on the trends found across the physical properties in your series.

STEP 4 Compare and contrast the general trends (eg increase, decrease or no change) in the physical properties, when there is a slight change in form across a homologous series.

GLOBAL CONTEXTS
Scientific and technical innovation

ATL SKILLS
Information literacy
Access information to be informed and inform others.

Polymers

You have seen that organic compounds can have long chains in their structure. Polymers are organic molecules that can have extremely long chains by bonding many individual units called monomers into a single larger unit. Natural polymers include hair, fingernails, wool, DNA and proteins. Synthetic polymers are common commercial compounds and include nylon, polyester, polyethylene, Teflon™ and Gore-Tex™. Chemists are now able to manipulate the structure of polymers in order to produce "designer" polymers to achieve a desired function. Different polymers with different properties and functions can be developed by starting with different monomers.

 Activity 8   **Research and development of a polymer**

In this activity, you will investigate how polymers are produced by making a "bouncing ball" polymer by combining white glue and borax (also known as sodium tetraborate—a component of many detergents). The monomer unit in the bouncing ball polymer is polyvinyl acetate, which is found in solution in the white glue.

The diagram below shows a monomer of polyvinyl acetate.

When borax $Na_2B_4O_7$ is dissolved in water, some borate $B(OH)_4^-$ ions form. The form of the monomer allows weak bonds to be made between the monomer and borax ions. Bridges are formed, connecting the polymer chains to one another. This is called a cross-linked polymer, as shown below. The cross-linking produces a 3D polymer with many open spaces for water to occupy.

(a) two molecules of PVA before cross linking

(b) after cross linking

The polymer made in this activity has some properties of a liquid—if left untouched, it will begin to lose its shape. But it also has some properties of a solid—it bounces and it shatters. Substances that exhibit characteristics of both solids and liquids are called non-Newtonian fluids.

As a team, you must design and carry out a procedure to develop a polymer to make the best bouncing ball (ie the ball that will bounce the highest when dropped from a height of 1.0 m), within a set budget. If there is a tie in maximum bounce height, the best bouncing ball will be the ball that has the lowest height/cost ratio.

Each team will be allowed to spend a total of 7.50 chembits on their procedure. The cost breakdown of available supplies is:

- 1 cm³ of glue = 0.10 chembits
- 5 g of borax = 0.15 chembits
- rental of beakers = 0.25 chembits/h
- rental of stirring rod = 0.10 chembits/h
- rental of lab bench = 1.00 chembits/class
- rental of balance = 0.50 chembits/h
- rental of graduated cylinder = 0.50 chembits/h.

[SAFETY] Wear eye protection and avoid skin contact with the borax and glue. Wear disposable gloves when moulding and testing the bouncing balls. Do not take the balls out of the laboratory.

Questions

a) Discuss how placing a cost on lab materials changes your approach to the activity.
b) **Justify** why your group's bouncing ball can be classified as a polymer.
c) Do some research and find out some other common manufactured products that are made of polymers.

GLOBAL CONTEXTS
Scientific and technical innovation

ATL SKILLS
Collaboration
Encourage others to contribute.

Disposal of plastics

In the past 100 years, research and development in the area of polymer chemistry has allowed production of many new plastics with many different forms and properties. Modern, lightweight plastics used in consumer packaging and products have made our life easier, but much of this plastic is used only once and ends up in landfill sites. Many types of polymer are unreactive and do not decompose quickly, so plastic waste remains in the ground for hundreds of years.

One of the best ways to reduce the amount of plastic that goes to landfill sites is to recycle the plastic materials to make new plastic products. This involves either simply melting the plastic and moulding to a new shape or depolymerizing it to reform the monomers, which can then by polymerized again. Most modern plastics can be recycled.

However, sorting the different types of plastic is costly and time-consuming. Recycling plastics requires much less energy than it takes to originally make the plastics. It also conserves the crude oil that synthetic polymers are made from.

Reflection on Topic 2

Organic chemistry is one of the fastest-growing areas of chemistry. There are applications in pharmaceuticals (new medicines and drugs), in materials science (new materials such as alloys and composites) and in biochemistry (the study of chemicals produced by the body). In all of these areas, understanding the form or structure of molecules is important to understanding their function. Chemists can now work backwards from the function of a substance they wish to have. They design it with computer modelling and then create the substance in the laboratory.

- How could knowledge of organic chemistry allow chemists to work backwards to design organic molecules that require a specific function? (That is, they start with an end function and work backwards to determine the materials to start with.)

- What responsibilities do chemists have in producing new organic molecules whose impact on the environment is minimal? What would happen if they developed an extremely useful organic molecule that could not be recycled and whose waste would result in a negative impact on the environment?

TOPIC 3

Forms of energy

Energy has many different forms and can be stored in many different forms. All of these forms of energy measure the ability of a body to do work on another body. All forms of energy are measured in joules (J). These forms of energy include chemical energy, thermal energy, potential energy, kinetic energy, nuclear energy, light energy and sound energy.

Stores of energy

Kinetic energy is the energy associated with a moving object. For example, a flywheel (a very heavy wheel) stores kinetic energy once it is moving. The potter's wheel is a flywheel. By kicking the bottom wheel with your foot, you set the pottery wheel spinning. When it reaches the desired speed, you can stop kicking and the wheel keeps spinning as energy is stored in the wheel (Figure 11.4). The wheel will eventually slow down as the force of friction transforms kinetic energy into thermal energy, which is dissipated to the environment.

TAKE ACTION

Investigate plastic recycling in your area (Figure 11.3). Does your community have a plastic recycling plan? Does your school have a plastic recycling plan?

Figure 11.3 Recycling of plastic is an important way to reduce waste in landfills and to conserve raw materials

PLASTIC

WEB LINKS

The problem of plastic pollution is increasing as evidenced by the formation of plastic gyres in our oceans. Search for midway message from the gyre at www.youtube.com for more information.

Figure 11.4 A potter's wheel takes a lot of force to spin around, but then keeps spinning

Gravitational potential energy (GPE) is the energy stored by an object due to its vertical position in a gravitational field. The higher an object is, the greater its GPE. Hydroelectric power stations store energy as GPE of the water in a high reservoir. This stored energy is transformed to kinetic energy when the water is released and falls. The kinetic energy of the falling water is then used to move a turbine that generates electricity.

Elastic potential energy is the energy stored in an object when it is stretched or squashed, such as when a catapult is stretched but not released.

Chemical energy is the energy stored in the chemicals in fuels, food and batteries. The chemical energy stored in food allows our body to stay alive, to grow and to move around. The chemical energy in a fuel such as coal or oil is transformed to thermal energy when the fuel is burned. Fossil-fuel power stations work by using the kinetic energy of steam (obtained from burning coal, oil or gas) to move a turbine that generates electricity.

Nuclear energy is stored in the Sun, and nuclear fuels such as uranium. Nuclear energy can be released through a nuclear reaction—nuclear fission or nuclear fusion.

Work and energy changes

Work is done when energy is transferred from one form to another. The work done is equivalent to the amount of energy transferred, in joules.

When a force is applied to a body and the body changes its motion in the same direction as the force, we say that work is being done on it. An energy change takes place. For example, when a car brakes, work is done by the brakes on the car to stop it moving. The work done by the brakes, using the frictional forces between the brake pads and the rotating wheels, equals the amount of kinetic energy lost by the car.

The law of conservation of energy states that in a closed system, energy can neither be created or destroyed, it can only be changed from one form to another. The initial kinetic energy of the car is transformed to other forms of energy, mostly sound and thermal energy.

Work is also done when an object is lifted against the gravitational pull of the Earth. The work done by lifting the object equals the amount of gravitational potential energy gained by the object.

Energy changes in a pendulum

In 1582, the young Galileo Galilei observed a lamp swinging back and forth in Pisa cathedral. Using his pulse to measure the swing, he found that at small angles the lamp oscillated with a constant time period. These early observations led to his invention of using a pendulum to regulate a mechanical clock. Galileo realized that for the pendulum to continue to swing backwards and forwards, the energy must be conserved, but how does it do this?

⊂⊃ CHAPTER LINKS
Chapter 5 on energy has examples of work done (eg in a wind turbine, or in an endothermic chemical reaction to break bonds in the reactants). Chapters 5 and 7 both describe the law of conservation of energy.

To understand how a pendulum conserves its energy, we must consider two different forms of energy: gravitational potential energy and kinetic energy (KE). As objects are raised higher from the ground, they have more GPE, according to the formula

$$GPE = mgh$$

where m is mass in kilograms (kg), g is the gravitational field strength (9.81 N/kg) and h is the height of the object above a defined reference point (m).

An object gains KE as its velocity increases according to the formula

$$KE = \frac{1}{2} mv^2$$

where m is mass in kilograms (kg) and v is the velocity of the object in metres per second (m/s).

According to the law of conservation of energy, energy may change form but the total amount of energy does not change. Energy cannot be created or destroyed, only transformed from one form to another. As the pendulum swings, GPE is transformed into KE and back again. If no energy is lost due to air resistance, the total energy of the pendulum is conserved. This means that the pendulum will continue to swing forever as energy is transformed from KE to GPE and back again. If there is air resistance, a tiny amount of KE is transformed into thermal energy on each swing, so the gain in GPE is reduced by the same amount on each swing (Figure 11.5).

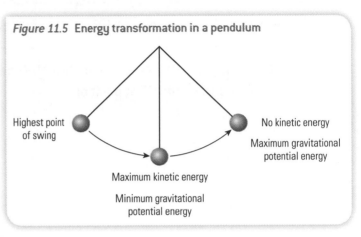

Figure 11.5 Energy transformation in a pendulum

Highest point of swing

No kinetic energy
Maximum gravitational potential energy

Maximum kinetic energy

Minimum gravitational potential energy

Activity 9 — Energy changes in a pendulum

In this activity, you will measure the actual velocity of a pendulum at the lowest point of its swing, and compare to the theoretical velocity (calculated from GPE lost = KE gained).

STEP 1 — Collect the following materials:

- stopwatch
- tape (not clear tape)
- about 3 m of fishing wire or string
- metre rule
- weight (to tie to the string)
- top pan balance.

STEP 2 — Read through Steps 3–11 and construct a data table to record the data you are going to collect and process.

STEP 3 **Measure** and record the mass of the pendulum weight. If the scale reading is in grams, divide by 1,000 to find the mass in kg.

STEP 4 Tie the string or fishing line to a beam on the ceiling, about 10 cm away from a wall. Make the length of the string as long as possible but make sure the weight will have enough room to swing freely.

STEP 5 Tie the other end of the string to the weight.

STEP 6 Pull the weight to one side, keeping the string tight, until the angle from the vertical is about 20°. Mark this point (position A) with a piece of tape and measure the vertical height of the weight relative to its starting position (position B, equilibrium position) when it is hanging straight.

STEP 7 Calculate and record the GPE of the weight when the pendulum is lifted to this height.

STEP 8 Calculate the theoretical velocity of the pendulum at position B if all the GPE at position A is transformed into KE at position B.

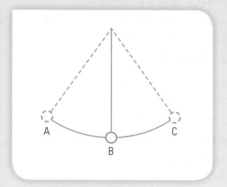

STEP 9 Release the pendulum at the measured height (from Step 6 above) and **measure** how long it takes to reach the maximum height on the other side (position C).

STEP 10 Repeat this measurement at least three times and find the mean in order to improve the accuracy of the result.

STEP 11 Calculate the average velocity throughout the swing. In order to calculate the velocity of the pendulum, you will need to use the equation

$$v = d/\Delta t$$

where v is average velocity of the object in metres per second (m/s), d is displacement in metres (m) and Δt is the change in time in seconds (s). In this case, d is the distance between the start and end points of the pendulum swing and Δt is the time for one swing from A to C.

TIP

Use pieces of tape on the wall to mark the height at which you will release the pendulum, and position C. The person using the stopwatch should make sure that her or his eye is parallel with the tape marker. This will reduce **parallax error** when judging when the pendulum reaches this position.

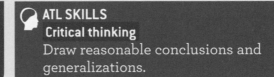
INTERDISCIPLINARY LINKS

You might be studying trigonometry in mathematics. You can find the actual distance travelled by the pendulum weight (the arc length) instead of measuring the horizontal displacement, by measuring the angle of the pendulum from the vertical. Use trigonometry to help you calculate the arc length AB in the diagram above, and use this in your calculation of the average velocity.

Questions

a) How similar were the theoretical and measured velocities?
b) Calculate the percentage difference.
c) Suggest reasons for the difference, using scientific reasoning.
d) Suggest improvements to the method and explain why these would reduce the difference between the theoretical and measured velocities.

GLOBAL CONTEXTS
Scientific and technical innovation

ATL SKILLS
Critical thinking
Draw reasonable conclusions and generalizations.

 **Activity 10**   Investigating a factor affecting pendulum swing

You are going to design an experiment to investigate one factor affecting the time of a pendulum swing. There are many aspects of the experiment that will need to be controlled. Brainstorm this with a partner before creating your inquiry question and procedure.

STEP 1   Choose one variable to investigate. Explain why you think this variable will affect the time of a pendulum swing.

STEP 2   Describe a detailed method for how you will test this hypothesis. Describe how you will alter the variables and collect data.
State the:
- independent variable (the variable you will change)
- dependent variable (what you will measure)
- control variables (all other variables that must stay the same during trials)
- materials (list the materials and equipment you will need to test your hypothesis)
- detailed method (so that anyone can repeat your experiment).

STEP 3 Construct a data table to record the data that you will collect. Include the units for the measurements.

Plot your data on an appropriate graph to show the relationship between the variable you investigated and the time of a pendulum swing.

Conclusion and evaluation

a) Describe what your data and graph show.

b) Does your conclusion agree with your hypothesis?

c) Evaluate the procedure and describe what improvements you would make if you were to carry it out again.

 GLOBAL CONTEXTS
Scientific and technical innovation

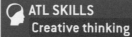 **ATL SKILLS**
Creative thinking
Use brainstorming and visual diagrams to generate new ideas and inquiries.

 **WEB LINKS**
Search for simulations where you can try your hand at designing a roller coaster.

Roller coasters

Roller coasters are exciting, but have you ever thought about how they work?

The law of conservation of energy is involved in all such rides. At the top of the first hill, the roller coaster possesses a high amount of GPE. As the roller coaster drops, the speed increases, reaching maximum speed at the bottom of the first hill. As the ride continues, KE and GPE continue to transform from one to another, although not perfectly, since energy is lost through friction, sound and heat. In order to give the roller coaster enough energy to continue to the end, the height of the first hill must be larger than any subsequent hills on the ride.

WEB LINKS
To try your hand at designing a skateboard course that keeps the rider safe and active, go to phet.colorado.edu/en/simulation/energy-skate-park.

Reflection on Topic 3

Kinetic and potential energy can be converted from one to another during pendulum motion, or on a roller-coaster ride. We know that power stations use energy resources such as nuclear energy, chemical energy or gravitational potential energy to generate electricity, which is then used to produce light and thermal energy for our homes.

- Where does energy come from?
- Is our supply of energy infinite?

Summary

You have observed the features of organisms, molecules or materials and a moving pendulum. Scientists often search for the connections between form and function.

In biology, you have seen examples where the form of an organism reflects its function. Can you think of any other examples in biology of specialized structures that are adapted to perform a specific function?

Often the different forms of an object or system can be categorized (eg the different forms of energy, the different states of matter or the different forms of organic molecules). Classification schemes like these help us identify patterns in form, which, in conjunction with the predicted function, can be used by research chemists to design new molecules with a specific function.

12 Movement

INQUIRY QUESTIONS

TOPIC 1 Movement in plants and animals

- **What stimuli trigger a movement response in woodlice?**
- **Can data support that movement in basking sharks is directional?**
- **Does the mechanical response of sensitive plants vary according to the stimulus?**
- **In what ways do the shape and size of winged seeds affect flight time and distance?**

TOPIC 2 Electrochemistry and movement of electrons

- **Why do electrons sometimes move spontaneously from one substance to another?**
- **What factors influence the voltage that a voltaic cell can produce?**

TOPIC 3 Velocity and acceleration

- **How can the velocity and acceleration of a moving object be determined?**
- **What causes the speed of an object to change?**
- **How does changing the angle of a ramp affect acceleration down the ramp?**

SKILLS

ATL

✓ Apply existing knowledge to generate new ideas, products or processes.

✓ Interpret data.

✓ Select and use technology effectively and productively.

✓ Draw reasonable conclusions and generalizations.

✓ Collect and analyse data to identify solutions and make informed decisions.

✓ Process data and report results.

✓ Gather and organize relevant information to formulate an argument.

✓ Analyse complex concepts and projects into their constituent parts and synthesize them to create new understanding.

Sciences

✓ Formulate a testable hypothesis and explain it using scientific reasoning.

✓ Design a method for testing a hypothesis, explaining how to manipulate the variables and how enough data will be collected.

GLOSSARY

Source of error a problem with an experimental method, the apparatus or ways in which the apparatus is used that cause measurements to be different from the true value.

Validity of the method refers to whether the method allows for the collection of sufficient valid data to answer the question. This includes factors such as whether the measuring instrument measures what it is supposed to measure, the conditions of the experiment and the manipulation of variables (fair testing).

COMMAND TERMS

Calculate obtain a numerical answer showing the relevant stages in the working.

Comment give a judgment based on a given statement or result of a calculation.

Select choose from a list or group.

✓ Organise and present data in tables ready for processing.

✓ Calculate the gradient of a straight-line graph.

✓ Interpret data gained from scientific investigations and explain the results using scientific reasoning.

✓ Plot scatter graphs with a line of best fit and identify trends.

✓ Describe improvements to a method, to reduce sources of error.

OTHER RELATED CONCEPTS

Consequences   Evidence   Interaction   Patterns

Introducing movement

Movement can be an action (eg the transfer of nutrients into a cell) or a change in position (eg when electrons are transferred between atoms or ions). Position changes in physics are usually defined with reference to a fixed starting point, or there can be problems with different observers saying different things about the motion of the same object.

Muscles, tendons, ligaments and bones are responsible for physical movement of the human body. Muscle cells generate contractive forces as they decrease in length and increase in width. Tendons transfer the forces exerted by these muscles to the bones. Bones support the body and maintain its shape, while also acting as levers to magnify small movements into larger ones. Joints controlled by a pair of muscles are able to convert the contractions of the muscles into both pushes and pulls.

This system for generating and controlling physical movement of the human body is strong and reliable, but accidents can happen. We put a lot of stresses on our bodies, particularly on joints. It is unusual to live into adulthood without some injury that has caused difficulty moving around. Injuries are particularly prevalent in the world of sport because sportspeople and athletes put even more stress on their bodies than others.

Baseball pitchers are particularly susceptible to injuries in an elbow ligament (that is, soft tissue that stabilizes the elbow). When this ligament is torn, the pitcher can no longer throw with any power. This injury used to mean the end of a pitcher's career. Then a doctor named Frank Jobe had an idea about how to repair the ligament (Figure 12.1) He removed a non-essential tendon from elsewhere in the body and threaded it through holes drilled into the bones of the upper arm and forearm, effectively stabilizing the inner side of the elbow.

Figure 12.1 The surgeon Frank Jobe performed the first ever "Tommy John surgery"

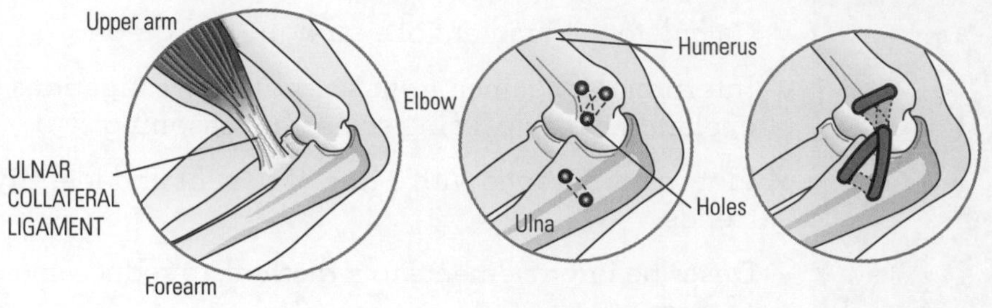

In March 2014, Dr Jobe died at the age of 88. Among the many tributes to him was one from the first pitcher whose career he had saved in 1974 with his first repair surgery. At the time, the player was told there was a one in 100 chance of success, but he went on to play for 14 more seasons and win over 160 more games. The operation that Dr Jobe pioneered has now been performed on almost 500 players and is referred to by the name of that first patient—Tommy John.

In this chapter, you will look at the motion of objects great and small, from sharks to maple tree seeds, and from ions in solution to decelerating cars. You will be considering the "why" or measuring the "how" of movement.

TOPIC 1

Movement in plants and animals

Movement is the process of changing from one place or position to another. In animals, movement can be triggered by a stimulus. If the response of the organism is to move toward or away from the stimulus, then the movement is referred to as taxis. If the movement is not in a particular direction as a result of a stimulus, then the movement is referred to as kinesis.

In plants, movement in response to environmental stimuli is referred to as tropism. Movement of abiotic factors can affect living things. Examples include the diffusion of materials across membranes, the movement of ocean currents containing nutrients, the dispersal of seeds by animals and wind, and the capillary action of water in soils.

In addition to the range of causes of motion, the mechanisms of motion can be studied.

Organisms may undertake migration from areas where conditions do not support the needs of the population to more favourable habitats. Human populations may migrate on a periodic basis (eg nomadic cultures following herds) or more permanently (eg in response to famine or conflict or limited economic opportunities).

Activity 1   Kinesis-investigating woodlice motion in response to a stimulus

The woodlouse is a common land-living crustacean that is easy to find in many locations around the world. They tend to inhabit moist, dark habitats such as in rotting vegetation under the surface of fallen leaves.

It is possible to test environmental factors that affect the motion of these organisms using a choice chamber that has different conditions on different sides. A container such as a Petri dish can be used, with a number of animals (eg eight individuals) in initially a central position.

[SAFETY] If you collect woodlice and handle them, wash your hands after doing so.

STEP 1   Formulate a hypothesis to test relating to the movement of the woodlice and damp/dry conditions, or light/dark conditions or the presence/absence of a substance you can find and which is ethical to use. State a dependent variable and an independent variable, and say what the effect will be and why.

STEP 2   Design a procedure to test the hypothesis.

Here are some questions to consider.

a) How will you compare and record the movement of the woodlice?
b) What variables will you control?
c) What ethical issues are there with this experiment, and how will you plan to resolve these issues?

STEP 3   Construct a results table to record your observations.

STEP 4   Describe your results, and **comment** on the **validity of the method** in this experiment.

🌐 **GLOBAL CONTEXTS**
Scientific and technical innovation

🧠 **ATL SKILLS**
Creative thinking
Apply existing knowledge to generate new ideas, products or processes.

Activity 2   Kinesis-analysis of data of motion in response to a food source

Basking sharks (*Cetorhinus maximus*) filter-feed on zooplankton (small floating marine animals) in temperate coastal seas. Marine biologists recorded the swimming paths taken by two basking sharks about 8 kilometres off the coast of Plymouth (UK). At the same time, the densities of zooplankton (in g/m^3) were recorded within 3 metres of the swimming path of the sharks.

a) Using the scale given, **calculate** the straight line distance
   i) from point A to point B
   ii) from point C to point D.
b) Outline the difference in the swimming behaviour between shark 1 and shark 2.
c) Using the data given, suggest reasons for the difference in the swimming behaviour of the two sharks.
d) Suggest two factors other than food that may affect the distribution of the basking sharks.

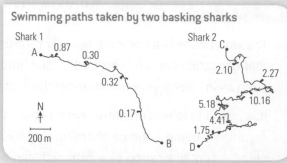
Swimming paths taken by two basking sharks

🌐 **GLOBAL CONTEXTS**
Scientific and technical innovation

💭 **ATL SKILLS**
Critical thinking
Interpret data.

🔗 **CHAPTER LINKS**
Chapter 15 on environment has further information on plant responses.

Many people would consider plants to be static—lacking the systems for movement that we see in animals, growing in one place and fixed in the soil. But plants show a number of remarkable features that allow them to move, sometimes quickly and sometimes slowly, in response to a stimulus.

👤 **Activity 3**   Investigating thigmotropism in the sensitive plant

Thigmotropism is the response of a plant to mechanical stimulation from an element in its environment. *Mimosa pudica* is known as the sensitive plant. This is because touching the plant causes the leaves to move from flat to folding inwards. A number of investigations of the sensitive plant are possible (eg is closing affected by the direction of the touch, the intensity of the touch, the frequency of the touch).

TIP
Videos of the closing can be taken and viewed frame by frame to analyse the closing and hence to suggest possible research questions.

Design an investigation you could carry out to measure the effect of one of the factors mentioned above on an aspect of thigmotropism. Describe which variables you will measure and control, and how you will collect data.

🌐 **GLOBAL CONTEXTS**
Scientific and technical innovation

💭 **ATL SKILLS**
Organization
Select and use technology effectively and productively.

Seed dispersal is important in the survival of plants as it minimizes competition between parent plants and offspring. Plants have evolved many methods, both physical and biological, by which to disperse their seeds.

Winged Acer seed

Maple trees (*Acer* sp.) produce seeds inside fruits that look like a pair of wings. These are carried by the wind as they fall from the tree. In a student experiment, 50 winged maple seeds were dropped, one at a time, from two different heights (0.54 m and 10.8 m). The histograms below show the distribution of the distance the maple seeds travelled.

a) For each height, identify the distance travelled by the greatest number of seeds.

b) Deduce the effect of height on seed dispersal.

c) Suggest two reasons for the effect of the drop height on the distance travelled by the seeds.

Student experiment: Maple seed dispersal

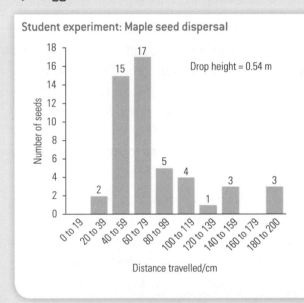

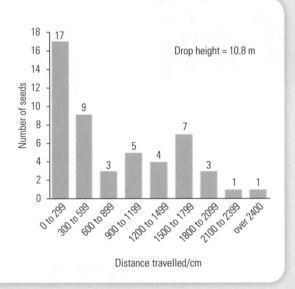

GLOBAL CONTEXTS
Scientific and technical innovation

ATL SKILLS
Critical thinking
Interpret data.

Reflection on Topic 1

Movement is the process of changing from one place or position to another.

- How do environmental conditions affect the reasons organisms change position?

- How would you convince someone that plants can move?

- What are the various mechanisms organisms use to move?

- How do conditions within the organism affect its motion?

Electrochemistry and movement of electrons

So far, throughout your study of chemistry, you have seen that there are many different types of chemical reaction. For example, the reaction of two substances could result in the formation of a precipitate or a gas. When an acid and a base react together, a neutralization reaction occurs as a result of a transfer of a proton (a hydrogen ion, H^+).

Another important type of chemical reaction is the oxidation–reduction type of reaction, in which electrons are transferred between two different substances. In oxidation reactions, electrons are lost. In reduction reactions, electrons are gained. The study of this movement of electrons is called electrochemistry.

Because these reactions always include both oxidation and reduction reactions, they are also called redox reactions. You are probably already familiar with many redox reactions—photosynthesis, corrosion (rusting) of iron and steel, spectacles that automatically darken in bright light, combustion of fuels, bleaching of fabric, processing of black-and-white photographic prints. All these examples are the result of the transfer of electrons.

One of the most common applications of redox reactions is in batteries. Batteries are used to provide an electric current to mobile phones, handheld game consoles, laptop computers, cars and smoke detectors. When electrons are transferred in a battery, an electric current is produced that can then do work in a circuit (eg to drive a motor).

In this section, you will be taking a closer look at the construction of batteries. The correct scientific name for a battery is a voltaic or electrochemical cell.

Activity series in metals

Why does the movement of electrons from one substance to another occur in redox reactions? Sometimes, when two substances react, one substance is more reactive (loses its electrons more easily) while the other substance is less reactive (gains electrons more easily).

Look at the following ionic equation for the reaction between zinc and a solution of a copper salt. You will see that zinc is more reactive than copper because the zinc metal has lost two electrons to become the zinc ion, Zn^{2+}. At the same time, the copper ion Cu^{2+} has gained two electrons to form copper metal. Two electrons have moved from the zinc metal to the copper ion. Zinc and copper are neutral atoms and therefore have no charge.

$$Zn(s) + Cu^{2+}(aq) \rightarrow Zn^{2+}(aq) + Cu(s)$$

TIP

Mnemonics:

OIL RIG—**o**xidation **i**s **l**oss, **r**eduction **i**s **g**ain.

LEO the lion says GER—**l**osing **e**lectrons is **o**xidation, **g**aining **e**lectrons is **r**eduction.

In this example, the zinc has oxidized (lost electrons) and the copper ion has been reduced (gained electrons). A redox reaction will only occur if a more reactive metal is placed in a solution of a less reactive metal ion. The reverse reaction will not occur because copper is less reactive than zinc.

$$Cu(s) + Zn^{2+}(aq) \rightarrow \text{no reaction}$$

The ranking of reactivity of metals is called an activity series.

Activity 5 — Activity series of metals

You are going to develop an activity series by investigating a series of reactions involving metals in aqueous solutions, using an online simulation. By interpreting your observations you will be able to organize a list of metals from the most reactive to least reactive.

Here are some key questions to ask during this activity.

- What is the relationship between the reactivity of a metal and its reactivity as a metal ion?
- What use would the information in an activity series be to chemists?
- Metals such as gold and copper occur naturally in the environment, while other metals such as aluminium and sodium only occur in their ionic form in salts. Investigate why this is the case.
- How could the information in the activity series be used to help mining companies determine how to extract metals that only occur as salts?

WEB LINKS
Try the simulation at www. tinyurl.com/3bgkay6 or just search for "simulation" and "metals in aqueous solutions".

GLOBAL CONTEXTS
Scientific and technical innovation

ATL SKILLS
Critical thinking
Draw reasonable conclusions and generalizations.

Electrochemical cells

An oxidation–reduction reaction can also be thought of as occurring in two half-reactions.

$$Zn(s) + Cu^{2+}(aq) \rightarrow Zn^{2+}(aq) + Cu(s)$$

$Zn(s) \rightarrow Zn^{2+}(aq) + 2e^-$	(oxidation half-reaction)
$Cu^{2+}(aq) + 2e^- \rightarrow Cu(s)$	(reduction half-reaction)

In some redox reactions, the movement of electrons is spontaneous and does not need an external energy source (eg battery or DC power supply). In other reactions, the movement of electrons is non-spontaneous and needs a source of electricity to make it happen (eg recharging a rechargeable battery). A voltaic cell is an electrochemical device that produces an electric current in a spontaneous redox reaction.

TIP

Mnemonics:
FAT CAT—(direction of electron flow) from anode to cathode.
ANO—anode, negatively charged and oxidation.
CPR—cathode, positively charged and reduction.

Figure 12.2 A model of an electrochemical cell

A voltaic cell consists of two metal electrodes (eg one zinc and one copper) that each dip into a solution of its own metal ion (called an electrolyte) as shown in Figure 12.2. There are always two sets of electrodes–electrolytes representing the two half-reactions; these are called the two half-cells. Unlike most other reactions (which take place in a single test tube or beaker), redox reactions in a voltaic cell always take place in two separate half-cells.

The two half-cells are connected by a conducting wire between the two electrodes. This wire acts to complete the circuit and allow the electrons to move from one half-cell to the other. A voltaic cell can also be connected to a voltmeter whose readings give the cell voltage, measured in volts. You will see from the diagram that the two half-cells are also connected by a salt bridge, which is a type of internal circuit.

The anode is the electrode where oxidation takes place. It is negatively charged as electrons are produced here. The cathode is the electrode where reduction takes place. It is positively charged. In a voltaic cell, the direction of electron movement is always from the anode to the cathode.

QUICK THINK
Look at Figure 12.2. Why is a salt bridge necessary?

Activity 6   Creating the best voltaic cell

You are going to carry out a series of trials to investigate the variables that affect the voltage produced by a given cell. You will then design an investigation to test the effect of one variable.

[SAFETY] Wear eye protection. Avoid skin contact with the copper sulfate and zinc sulfate solutions.

STEP 1 Based on Figure 12.2 and the equipment, electrodes and electrolytes provided by your teacher, build a voltaic cell.

STEP 2 Investigate the conditions that affect the voltage produced by this cell.

STEP 3 Once you have carried out trials on this electrochemical cell, **select** one variable to manipulate. Design an investigation into how this variable affects the voltage produced. List the variables that must be controlled.

STEP 4 What conditions must be in place to produce the maximum voltage?

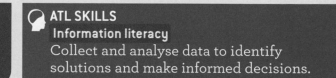

🌐 **GLOBAL CONTEXTS**
Scientific and technical innovation

🧠 **ATL SKILLS**
Information literacy
Collect and analyse data to identify solutions and make informed decisions.

Fuel cells

New research in electrochemistry has led to the development of fuel cells. Fuel cells are similar to electrochemical cells, but require that the reactants (fuel and air or oxygen) are constantly supplied to give the cell a long operating life. In the simplest fuel cell, hydrogen is the fuel and the only product formed is water. The chemical reactions that take place in fuel cells are redox reactions and involve movement of electrons.

Governments, private industry and research institutions are collaborating to investigate fuel cells. Because fuel cells are very fuel-efficient when compared with combustion engines and do not produce greenhouse gas emissions, they could be the future for cars. Automobile manufacturing companies are currently developing mass-market cars that have fuel cells as their source of energy. Many countries such as Germany and Norway are setting up a hydrogen-refuelling infrastructure, although this will take many years to be comparable with the distribution of petrol stations.

Reflection on Topic 2

- How has the science behind the movement of electrons developed into the field of electrochemistry?

- Do you think rechargeable batteries have less environmental impact than non-rechargeable ones?

- How can you make better and more sustainable decisions about battery use in your personal choices and in your community?

⚡ TAKE ACTION

Batteries, both rechargeable and non-rechargeable, contain toxic substances such as cadmium, lead, mercury, copper, zinc, manganese, lithium and potassium.

Does your school or community currently have battery drop-off sites so that these batteries can safely be disposed of?

If so, how can you promote the use of these sites to your school community?

If not, what plan could you devise to ensure that a battery-recycling programme is established in your community?

TOPIC 3

Velocity and acceleration

In physics, a major area of study is how and why objects move the way they do. Using formulae derived from scientific laws such as Newton's laws of motion, it is possible to predict the motion of an object as it travels from one location to another.

The motion of objects can be measured by investigating the change in position (displacement), speed or acceleration of an object.

Velocity

This is the speed and direction of an object. Velocity is measured in metres per second (m/s). Since velocity is associated with a direction, a particular direction of travel must be set as positive (+). Thus, it can be seen clearly if an object begins to move in another direction.

Consider this example. Juan is driving north at a speed of 60 km/h or 16.7 m/s. He stops at the shop for a few minutes, then drives back south at a speed of 40 km/h or 11.1 m/s. Describe the velocity of Juan during his trip.

Answer: North = positive direction (+)

Juan driving north, velocity = +16.7 m/s

Juan at the store, velocity = 0 m/s

Juan driving south, velocity = −11.1 m/s

QUICK THINK

Comment on the motion and velocity of the object by examining the displacement–time graphs in Figure 12.3.

Velocity is the time it takes for an object to travel a certain displacement (distance in a certain direction). It can also be determined by finding the slope or gradient of the line in a displacement–time graph.

velocity (m/s) = **displacement (m) / change in time (s)**

Vectors are quantities that have direction as well as a magnitude (size). Velocity is a vector.

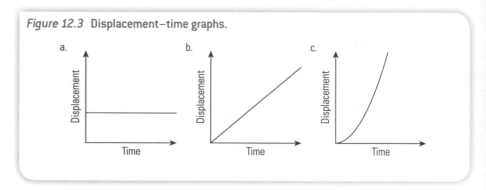

Figure 12.3 Displacement–time graphs.

Acceleration

Acceleration is the rate of change of velocity. It is measured in metres per second per second (m/s²). Since it involves velocity, acceleration is also directional—acceleration is a vector. When acceleration occurs, it can mean a change in direction or change in speed. A negative acceleration means a decrease in speed or an acceleration in the opposite direction of travel; this is usually called deceleration.

$$\text{Acceleration (m/s}^2) = \frac{\text{change in velocity (m/s)}}{\text{change in time (s)}} = \frac{\text{final velocity} - \text{initial velocity}}{\text{change in time (s)}}$$

Consider this example. A high-speed train travelling at 200 km/h slows to stop at a station in 40 seconds. What is the acceleration of the train?

Answer: Initial velocity = 200 km/h

$$= \frac{200\,\text{km}}{1\,\text{h}} \times \frac{1000\,\text{m}}{1\,\text{km}} \times \frac{1\,\text{h}}{3600\,\text{s}} = 55.6 \text{ m/s}$$

Final velocity = 0 m/s

Time taken = 40 s

QUICK THINK

Comment on the motion and acceleration of the object by examining the velocity–time graphs in Figure 12.4.

$$\text{Acceleration} = \frac{0 - 55.6 \text{ m/s}}{40\,s} = 1.38 \text{ m/s}^2$$

Acceleration can also be determined by finding the slope or gradient of the line in a velocity vs time graph.

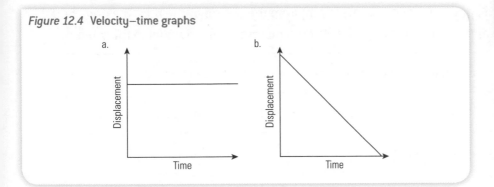

Figure 12.4 Velocity–time graphs

🔗 **INTERDISCIPLINARY LINKS**
You may also study velocity–time graphs in MYP mathematics.

Balanced and unbalanced forces

Another example of a vector is force. There are two different types of force that we will look at—balanced and unbalanced forces. When two forces are equal in size but work in opposite direction they are said to be balanced forces.

In contrast, when two forces are not equal in size they are said to be unbalanced forces.

The overall force acting on an object is called the resultant force. For balanced forces, this is zero. For unbalanced forces, the resultant force varies; it is equal to the difference between the two forces and acts in the direction of the greater force (Figure 12.5).

QUICK THINK

Search online for distance, velocity and acceleration data in sports that use metric units, such as rugby, athletics and swimming. Use the data to practise calculations, for example calculate and compare average velocity and acceleration in 100 m and 200 m races.

QUICK THINK

How is a parachute jump an example of unbalanced forces?

Figure 12.5 The resultant of two unbalanced forces

 Activity 7 Investigating forces

STEP 1 Collect the following materials:
- two protractors
- string
- iron ring
- two Newton meters.

STEP 2 Place two protractors back-to-back on the table to form a circle.

STEP 3 Connect two 10 cm pieces of string to a small iron ring.

STEP 4 Connect a Newton meter to the other end of each piece of string.

STEP 5 Two people per group will pull one Newton meter away from the centre of the ring. The ring should be kept over the centre of the protractors and the Newton meters pulled parallel to the table.

STEP 6 Measure and record the values of the two forces when the ring does not move.

STEP 7 Working together, pull the Newton meter with various forces. Observe and note what happens.

STEP 8 Sketch and label a vector diagram for each trial.

STEP 9 **Calculate** the resultant force for each trial.

Discussion

a) Discuss examples of where we see balanced and unbalanced forces in everyday life.

b) Give a scientific explanation of what happens in a car when the driver suddenly applies the brakes. Make sure that your explanation includes the term force.

🌐 **GLOBAL CONTEXTS**
Scientific and technical innovation

💭 **ATL SKILLS**
Information literacy
Process data and report results.

👥 Activity 8   Measuring velocity

You are going to investigate how changing the angle of the ramp changes the average velocity of a toy car as it rolls down the ramp.

[SAFETY] If heavy wooden ramps are used they should be placed where they will not easily fall. On the floor or on a large bench is best; not between two tables.

STEP 1 Locate the following materials:

- small toy car
- flat wooden board to use as a ramp—at least 1 m long
- books or blocks to elevate the ramp
- metre rule
- stopwatch
- protractor.

STEP 2 How do you expect the velocity of the car at the bottom of the ramp to change as the angle of the ramp changes? Give a scientific explanation for your prediction.

STEP 3 Read through steps 4–7 and state the:

- independent variable (the variable you will change)
- dependent variable (the variable for which transformed data will be collected)
- control variables (all other variables that must stay the same) for this experiment.

Construct a data table for the measurements you will take. Include units of measurement for each variable. Make sure you understand how the data will be transformed to find the car's average velocity.

STEP 4 Measure the length of the ramp; this is the distance the car will travel.

STEP 5 Using a book, elevate one end of the ramp. Measure the angle of the ramp from the table to the top surface of the ramp using the protractor.

STEP 6 Place the car at the top of the ramp, then measure and record the time it takes for the car to roll to the bottom of the ramp.

STEP 7 Repeat step 6 at least 3 times, and find the mean in order to improve the accuracy of the result.

STEP 8 **Calculate** the average velocity of the car in metres/second (m/s) at this ramp angle. Make sure that the distance used in the formula for average velocity is the length of the ramp.

STEP 9 Repeat steps 5–8 with more books to increase the angle of the ramp. Conduct the experiment for at least 5 ramp angles.

STEP 10 Plot a graph to show how the angle of the ramp affects the average velocity of the car.

Conclusion and evaluation

a) Describe how changing the angle of the ramp affects the average velocity of the car.
b) Was your hypothesis supported by your data? Explain using data from your analysis.
c) Suggest possible **sources of error** in this experiment.
d) Suggest how the procedure can be improved to prevent or minimize these errors.
e) To extend the investigation, instead of changing ramp angle, what other variables could be tested using a similar method? Formulate a hypothesis for this new test.

 **GLOBAL CONTEXTS**
Scientific and technical innovation

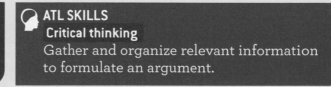 **ATL SKILLS**
Critical thinking
Gather and organize relevant information to formulate an argument.

You are going to investigate the change in acceleration as the angle of a ramp changes. How does changing the angle of the ramp affect acceleration?

[SAFETY] If heavy wooden ramps are used they should be placed where they will not easily fall. On the floor or on a large bench is best; not between two tables. Place something for catching the trolley at the bottom of the ramp.

STEP 1   Locate the following materials:

- ticker timer and power supply
- ticker tape
- trolley
- books or blocks to elevate the ramp
- metre rule
- protractor
- sticky tape to stick the ticker tape to the trolley.

STEP 2   State how you expect the acceleration of the car to change as the angle of the ramp changes. Explain why you think this will happen.

State the:

- independent variable (the variable you will change)
- dependent variable (the variable for which transformed data will be collected)
- control variables (all other variables that must stay the same).

STEP 3   Read through the steps 4–15 and construct a data table for the measurements you will take and the processed data. Include units of measurement for each variable. Make sure you understand how the data will be transformed to find the car's acceleration.

STEP 4   Using a book, elevate one end of the ramp. Measure the angle of the ramp from the table to the top surface of the ramp using the protractor.

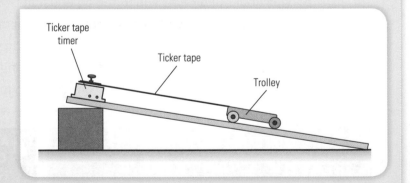

STEP 5   Place the trolley at the top of the ramp. Use tape to stick one end of the ticker tape to the trolley and thread the other end of the tape through the ticker timer.

STEP 6   Switch on the ticker tape timer and let the trolley run down the ramp.

STEP 7 Remove the ticker tape from the trolley. Cut the tape into sections of 6 dots, each with 5 spaces between dots as shown opposite.

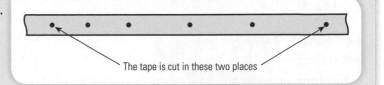

The tape is cut in these two places

STEP 8 If the timer produces 50 dots each second, **calculate** the time that 5 dot spaces (the length of each strip) represents.

STEP 9 Arrange successive ticker tape strips along a horizontal line (time axis) on a piece of paper. Glue the strips so that the bottom of each strip touches the axis. The complete chart should look like this:

STEP 10 Find the distance travelled by the trolley for the first and the last strips. This is equal to the length of the tape.

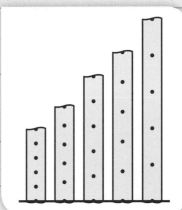

STEP 11 **Calculate** the average speed of the trolley during the first strip and during the last strip.

STEP 12 Find the time interval between the start of these strips.

STEP 13 To calculate the acceleration of the trolley between the first and last strips, use the equation:

acceleration = change in speed / change in time.

STEP 14 Repeat steps 4–13 with more books to increase ramp steepness. Conduct the experiment for at least 5 ramp angles.

STEP 15 Plot a graph to show how the angle of the ramp changes the acceleration of the car along the ramp.

Conclusion and evaluation

a) Describe how changing the angle of the ramp affects the acceleration of the car down the ramp.

b) Was your hypothesis supported by your data? Explain using numbers from your analysis.

c) Suggest possible sources of error in this experiment.

d) Suggest how the procedure could be improved to prevent or minimize these errors.

🌐 **GLOBAL CONTEXTS**
Scientific and technical innovation

💭 **ATL SKILLS**
Critical thinking
Analyse complex concepts and projects into their constituent parts and synthesize them to create new understanding.

Research and discuss the
activation of airbags in
a car; look specifically at
deceleration. How is airbag
release triggered?

Deceleration and airbags

Airbags are activated during collisions to protect the car occupants from direct impact with the car. This reduces injury. Today's cars may have up to eight airbags placed around the front and side of the vehicle.

Reflection on Topic 3

Throughout this topic you have investigated and discussed three different vectors: force, acceleration and velocity. You specifically looked at balanced and unbalanced forces at work in our world today, but how do these forces relate to Newton's third law of motion?

The second vector discussed was velocity, which is speed in a given direction. In activities 8 and 9 you determined the relationship between velocity and acceleration.

- Discuss how the gradient/slope of the velocity–time graph can be used to find the acceleration of the object.

Summary

You have looked at examples of the movement of animals and plants, electrons and vehicles. In plants and animals, movement can be a response to some stimulus from the environment. Animals may migrate from areas of unfavourable conditions. In chemistry, the reactions in which electrons move are called redox reactions because they involve both an oxidation reaction and a reduction reaction. In physics, study of motion is concerned with how and why objects move as they do—understanding velocity and acceleration.

The study of movement involves looking at both the mechansims and the causes of movement, which can be used to predict or model the motion of an object and how its motion changes.

Function

INQUIRY QUESTIONS

TOPIC 1 Function in organisms

- What variables affect two-point discrimination?
- How does colour distribution change as a fruit ripens?
- What characteristics affect our ability to complete a mirror test?

TOPIC 2 Function of acids and bases

- What is the function of an indicator in a chemical reaction?
- What are the relationships between acids and bases?
- Why is the process of neutralization an important idea in the fields of both chemistry and biology?

TOPIC 3 Function of the eye and spectacles

- How does the eye function?
- How do glasses or contact lenses help people with vision problems?

SKILLS

ATL

✓ Recognize unstated assumptions and bias.

✓ Test generalizations and conclusions.

✓ Apply skills and knowledge in unfamiliar situations.

✓ Interpret data.

✓ Collect and analyse data to identify solutions and make informed decisions.

✓ Gather and organize relevant information to formulate an argument.

✓ Revise understanding based on new information and evidence

✓ Make guesses, ask "what if" questions and generate testable hypotheses.

✓ Apply existing knowledge to generate new ideas, products or processes.

Sciences

✓ Formulate a testable hypothesis using scientific reasoning.

✓ Design a method to test a hypothesis, and select appropriate materials and equipment.

✓ Ensure sampling is random to avoid selection bias.

✓ Create accurate, labelled scientific drawings.

GLOSSARY

Random sampling a method to select samples randomly and entirely by chance.

Selection bias a type of bias caused by using non-random data.

Sensitivity the smallest change in value that an instrument can detect.

COMMAND TERMS

Calculate obtain a numerical answer showing the relevant stages in the working.

Evaluate make an appraisal by weighing up the strengths and limitations.

Explain give a detailed account including reasons or causes.

Formulate express precisely and systematically the relevant concept(s) or argument(s).

Identify provide an answer from a number of possibilities. Recognize and state briefly a distinguishing fact or feature.

✓ Interpret data gained from scientific investigations and explain the results using scientific reasoning.

✓ Describe improvements to a method, to reduce sources of error.

✓ Use appropriate scientific conventions to visually represent refraction.

✓ Use appropriate scientific terminology to make the meaning of your findings clear.

OTHER RELATED CONCEPTS	Form Evidence Patterns

Introducing function

Geckos are widely considered to be among the world's greatest climbers. These amazing lizards are capable of rapid movement on almost any surface and any angle. They captured Aristotle's attention more than 2,000 years ago as he noted "it can run up and down a tree in any way, even with the head downwards".

But exactly how does the gecko manage this trick? The simple explanation would be that geckos have sticky feet, but they do not. Some have speculated that geckos use tiny water droplets to create surface tension, or electrical charge that allows them to walk upside down on the ceiling.

Whatever the secret, it must be that the function of gecko toes or feet explains this ability. Function is defined as a purpose, a role, or a way of behaving that can be investigated. So scientists investigated gecko toes very closely to see how they functioned.

Gecko toes are remarkable (Figure 13.1). Each one is divided into small ridges called lamella, on which are tens of thousands of tiny hair-like structures called setae. The setae, about a tenth of the diameter of a human hair, are packed very tightly—there are almost 5,000 of them in every square millimetre.

But this is not the end of the story. Using an electron microscope to examine the setae reveals that each seta has hundreds of even smaller projections with flat tips (Figure 13.2).

These microscopic projections are able to take advantage of weak intermolecular forces called van der Waals forces that exist between the ends of the projections and molecules in the surface the gecko is climbing on. Normally, van der Waals forces are too weak to have any noticeable effect—they work only at very close range. But the tiny ends of the "hairs" on the gecko's feet make such close contact with surfaces that each interaction provides the maximum force. And given the

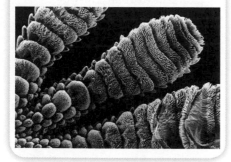

Figure 13.1 Close-up of a gecko's foot

Figure 13.2 Scanning electron microscope image of gecko foot hairs

enormous number of "hairs", the interactions add up to a total contact force so large that a gecko could hang upside down from any surface using only a single toe.

Could we design something to mimic the functionality of gecko feet? And that would have a practical use for humans?

Enter the gecko robot with a special footpad. These machines are still in the experimental phase but they are capable of climbing walls and ceilings. Some scientists are looking ahead to their potential functionality in space. Imagine an army of gecko robots able to climb over and stick to any surface. They could, for example, clean and repair the surfaces of spacecraft and satellites—orienting themselves on any surface in zero gravity and in a vacuum at freezing temperatures.

The idea of looking at the relationship between form and function in biology and applying it to our lives is known as biomimicry. This is a growing field of study. Velcro® mimics the function of burrs (seed cases that cling on with hooks or prickles). The structure of shark skin was studied to produce swimsuit materials that reduce drag in the water. One of the most important innovations in transport in the past two centuries is the jet plane; it functions like a bird's wing.

In this chapter, you will look at function in a number of settings. For example, you will examine the function of the nervous system; the function of indicators, acids and bases; and how the eye functions and spectacles help us see better.

TOPIC 1

Function in organisms

Organisms and their structures have purposes, roles or ways of behaving that can be investigated. These functions can occur at the level of molecules, cells, tissues, organs or organ system.

The nervous system's function is to control and coordinate the activities of other organ systems. These functions include:

- collection of information from sensory organs
- coordination of responses to external stimuli
- coordination of muscle activity
- monitoring the functioning of other organ systems
- regulation of the body's response to internal stimuli.

The two main divisions of the nervous system are the central nervous system (CNS) and the peripheral nervous system (PNS). The CNS is composed of the brain and the spinal cord. The PNS is composed of all the other nerve tissue that is not part of the central nervous

CHAPTER LINKS
Chapter 16 on balance has an analysis of how human body temperature is maintained by body systems, which include a structure in the brain called the hypothalamus.

system. Nerve cells are known as neurons. Nerves transmit sensory information to the CNS, for example from the skin or the eyes, via sensory neurons, while motor neurons conduct signals from the CNS to the muscles and the hormone system.

Each and every sensory neuron connects to the CNS. A given sensory neuron responds to all information from its input area, for example an area of the skin, as if it were coming from one point. This input area is called the receptive field of the neuron. On the shoulder, for example, each sensory neuron gathers information from a much larger skin area than a neuron on the fingertip (ie the receptive field is larger on the shoulder). In order for a person to feel two points, two separate sensory neurons must be activated by stimulation of their respective receptive fields.

 Activity 1 Two-point discrimination

The photo opposite shows a diagnostic tool used by neurologists to test for nerve damage. The tool is used to assess the patient's ability to discriminate between feeling one or two points of contact on the skin. This tool can be modelled by bending a paper clip into a U-shape so that the two tips are separated by 2 cm. Either one or both ends of the paper clip are placed on the skin. The person being tested must look away as the test is carried out, and says whether he or she can feel one or two points. If the points of the paper clip are touched to a fingertip, a person will sense it as two points. But if the two points are touched to the back of the shoulder, the person is likely to sense it as a single point.

a) Suggest why sensory neurons in different parts of the body have different receptive field areas.

b) Design an experiment to investigate whether two-point discrimination is fixed or whether it can be influenced by external variables.

TIP

There is potential for **selection bias** when choosing whether to apply one or two points. To control for bias, scientists use **random sampling**. To ensure that your subjects are unable to detect a pattern in your use of one point or two points, a random number generator or coin flipper could be used (heads to direct the application of one point and tails to direct the application of two points).

Questions to consider:

- How many times will you repeat the test in each area of skin?
- How large must the difference be in the results for two areas to give conclusive evidence of difference in two-point discrimination?

[SAFETY] Only apply gentle pressure.

WEB LINKS
Go to www.random.org/coins/ or search for "random coin flip".

🌐 **GLOBAL CONTEXTS**
Scientific and technical innovation

🧠 **ATL SKILLS**
Critical thinking
Recognize unstated assumptions and bias.

The function of sensory neurons is gathering information and passing it to the CNS. The function of the CNS is to process the information and send the impulses to generate a response. The function of a motor neuron is to carry information from the CNS to the body part whose function is to generate a response.

The processing of visual information is asymmetric: each half of the brain interprets visual information that is on the other side of the body.

 Activity 2   **Motor neurons and mirror drawing**

Mirror drawing requires individuals to create drawings while being guided by the reflection of an image they are attempting to copy or trace. This explores the connection between a fine motor task (where you need to move muscles precisely) and the role of sensory information in guiding the task. In the apparatus shown in the diagram opposite, subjects are asked to trace inside the lines between an inner and outer star.

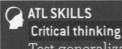

Mirror supported by report stand

Sheet of hardboard supported by report stand

Sheet of paper with inner and outer star

Drawing pin

Construct a table to record the number of the trial, the length of time to complete the trial and the number of errors made in the trial.

Design an experiment to investigate a variable that you predict will influence the success of mirror drawing. Be sure to state the basis of your prediction.

WEB LINKS
The mirror test can be carried out online. Search for "online psychology laboratory" and "mirror test" for one example.

 GLOBAL CONTEXTS
Scientific and technical innovation

ATL SKILLS
Critical thinking
Test generalizations and conclusions.

 Activity 3   **The function of fruit colour**

There is usually a relationship between fruit ripeness and colour. Unripe fruits are often dominated by the green pigment molecule chlorophyll. As the fruits ripen, a colour change occurs. The chlorophyll degrades and as green colour fades, other coloured pigments (that were hidden by the green colour) are revealed. At the same time, additional coloured pigments are produced.

TIP

It is possible to determine the relative amount of green, red and blue pigment in fruit skin by analysing photographs of the fruit. Smartphone and tablet photography or colour-picking apps allow determination of the reds, greens and blues (RGB values) reflected from a surface.

 STEP 1 Conduct research to address the question:
what function do fruit pigments perform in unripe and ripe fruit?

STEP 2 Design an experiment to investigate one of the following questions:

a) How do colour levels change during the ripening process?
b) How do environmental conditions affect colour changes?

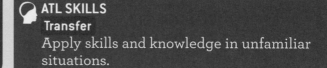

🌐 **GLOBAL CONTEXTS**
Scientific and technical innovation

🗣 **ATL SKILLS**
Transfer
Apply skills and knowledge in unfamiliar situations.

Reflection on Topic 1

The function of an organism or structure can be defined as the purpose, role or way of behaving that can be described and investigated. In order to more clearly understand function, it is useful to study the form of the structures that allow the function to happen.

- How does studying form enable us to better understand function?

- The fruit colour distribution experiment leads to the question: how is function in one organism affected by function in other organisms?

TOPIC 2

Function of acids and bases

Acids and bases are found widely in our everyday environment. Examples include the soap you use in the shower, the orange juice you drink for breakfast and the liquid used to clean windows. Whether a solution is an acid or a base can be detected through the use of an electronic pH meter that gives a numerical reading or an indicator that changes colour. Colour changes can be compared against a scale supplied with the indicator.

An indicator's main function is to identify whether or not a solution is acidic or basic. Knowing this has many practical functions. Indicators are used by gardeners to identify the pH of soil and so to identify which crops and plants will grow best in that soil. Lifeguards use indicators to establish the pH of water in a swimming pool. Meteorologists use indicators to identify the pH of precipitation to see if it is harmful to the environment. Indicators can be used in laboratories to help identify unknown substances.

Neutralization

Acids and bases neutralize each other's properties. When an acid is neutralized by a base, a hydrogen ion (H^+) from the acid reacts with a hydroxide ion (OH^-) from the base to form a molecule of water (H_2O):

acid + base → water + salt

$$HF + KOH \rightarrow H_2O + KF$$

Neutralization is the reaction between an acid and a base to form water and a salt. A solution is considered neutral when its pH is 7 (at room temperature). At really high temperatures, neutral solutions are slightly acidic and have a pH of about 6.8.

 **WEB LINKS**
To find out the importance of acids and bases and their pH levels in everyday life go to www.youtube.com and search for Importance of pH in everyday life.

 Activity 4   **Identification of substances**

Forensics is the study and identification of unknown substances from crime and accident scenes. Chemists also need to identify unknown substances. There are a variety of different methods using chemical tests or instruments to aid in this identification. The function of each test is to distinguish one substance from another.

In this activity, you will develop a scheme for classifying different substances as acids, bases or salts. This classification is made possible by examining the function of various indicators and reagents. Each indicator and reagent has a characteristic way of reacting with other substances. Once you have determined their characteristic reaction, you will use these indicators and reagents to identify unknown substances.

This activity uses a 96-well plate so that the experiments can be carried out on a microscale. Microscale experiments reduce waste, reduce costs and increase lab safety because small amounts of chemicals are needed.

STEP 1 Collect the following materials:

- eye protection
- microwell plate
- dilute hydrochloric acid
- magnesium ribbon
- sandpaper or steel wool
- conductivity meter

- distilled water
- blue and red litmus paper
- phenolphthalein
- bromothymol blue
- universal indicator solution

- solutions of: ammonium sulfate, ethanol (C_2H_5OH), hydrochloric acid and sodium hydroxide
- salts labelled A–D.

STEP 2 Create a matrix of solutions and reagents to be tested as shown in the diagram opposite.

	TEST SOLUTIONS						
T E S T R E A G E N T S		HCl	NaOH	H₂SO₄	HNO₃	KOH	Ba(OH)₂	H₂O
	Phenolphthalein	○	○	○	○	○	○	○
	Bromothymol blue	○	○	○	○	○	○	○
	Universal indicator	○	○	○	○	○	○	○
	Magnesium	○	○	○	○	○	○	○
	Calcium nitrate	○	○	○	○	○	○	○
	Conductivity	○	○	○	○	○	○	○
	Litmus paper	○	○	○	○	○	○	○

STEP 3 Construct a table to record all the necessary observations.

STEP 4 Add a drop or two of each test solution into the wells. Add a drop of each test reagent to the test solutions.

[SAFETY] Wear eye protection at all times. For all the test solutions apart from water, avoid contact with eyes, mouth and skin.

STEP 5 Use a piece of sandpaper or steel wool to clean a piece of magnesium ribbon until it is shiny. This removes the oxide coating so that the surface of the magnesium can react more easily. Add a small piece of magnesium to the test solutions.

STEP 6 Use the conductivity meter to test the solutions. Make sure you rinse the meter with distilled water after each test.

STEP 7 Use litmus paper to test the solutions.

STEP 8 Based on your observations, **formulate** a procedure that will allow you to classify your test solutions as acids, bases or salts.

You will now apply your procedure to identify unknown substances.

[SAFETY] Wear eye protection at all times while working with the unknown substances, and avoid skin contact with them all.

STEP 9 **Identify** the four different solutions that are labelled A–D. The possible identities of the four unknown solutions are: ammonium sulfate, ethanol (C_2H_5OH), hydrochloric acid and sodium hydroxide.

STEP 10 **Identify** the four different white crystals that are labelled A–D. Some of these are salts and some are not. The possible identities of the four unknown solids are: calcium chloride, calcium hydroxide, citric acid ($HC_6H_7O_7$) and glucose ($C_6H_{12}O_6$).

STEP 11 Based on your observations, describe the properties of acids, bases and salts.

TIP

When acids and bases dissolve in water, they ionize to form H^+ or OH^- ions respectively. However, it is incorrect to state that substances that contain an –H group are acidic and that substances containing an –OH group are basic.

Many substances that contain hydrogen groups are not acidic and many substances containing a hydroxide group are not bases. The set of organic chemicals classified as alcohols all contain an –OH group but do not ionize when dissolved in water. Table sugar (sucrose) $C_6H_{12}O_6$ also contains –H and –OH groups but when dissolved in water it does not act as an acid or a base. The hydrogen is not released as an H^+ ion and the OH group is not released as an OH^- ion.

🌐 **GLOBAL CONTEXTS**
Scientific and technical innovation

🧠 **ATL SKILLS**
Critical thinking
Test generalizations and conclusions.

Activity 5 — Neutralization

In this activity, you will perform a neutralization reaction between hydrochloric acid and sodium hydroxide (a base). You will then evaporate the water so that the salt, sodium chloride, remains.

STEP 1 Collect the following materials:

- eye protection
- dilute hydrochloric acid (1 M)
- sodium hydroxide solution (1 M)
- a pipette for each flask
- distilled water
- phenolphthalein solution
- red and blue litmus paper
- evaporating dish
- tongs
- hot plate
- 10 cm³ graduated cylinder.

[Safety] Wear eye protection at all times. Dilute hydrochloric acid and sodium hydroxide are irritants: avoid contact with eyes, mouth and skin. Do not taste the sodium chloride.

The indicator phenolphthalein changes colour close to the point that a solution turns from basic to acidic. In basic solutions, phenolphthalein is pink. In acidic solutions, it is colourless. This colour change during the test indicates that your solution is nearly neutral.

STEP 2 Measure 5.0 cm³ of distilled water in a graduated cylinder and then pour it into the evaporating dish.

STEP 3 To the evaporating dish add 10 drops of sodium hydroxide and 1 drop of phenolphthalein.

STEP 4 Swirl the dish slowly to mix the indicator and the base. Observe what happens.

STEP 5 Very slowly and carefully, add hydrochloric acid one drop at a time to the solution in the evaporating dish until the solution just changes to a clear colour. Stir after each drop of acid that you add.

STEP 6 Put one more drop of the base in the solution. Observe what happens. If there is no colour change, add an additional drop until it turns pink.

STEP 7 Repeat step 5. Put one drop of acid at a time into the solution until the colour turns clear.

STEP 8 Put the evaporating dish on the hot plate, heat very gently and allow the liquid to evaporate slowly. Do not overheat.

STEP 9 When the liquid is almost completely evaporated, remove the evaporating dish from the hot plate and allow the residual heat to complete the evaporation process.

STEP 10 Make observations of the crystals that form.

STEP 11 Put about 2 cm³ of distilled water into the evaporating dish and test the new solution with red and blue litmus paper. Record any colour changes. Is the new solution an acid, a base or neither?

Discussion questions

a) **Explain** the importance of the neutralization reaction in terms of work in the laboratory and in the natural environment.

b) State which acid and base you would have to react together in order to produce the salt sodium nitrate.

 GLOBAL CONTEXTS
Scientific and technical innovation

ATL SKILLS
Critical thinking
Interpret data.

 **Activity 6**   **Which antacid is the most effective?**

Antacids are common medications to treat the symptoms of acid indigestion. The function of an antacid is to neutralize the excess stomach acid that causes indigestion. Antacids may contain sodium hydrogen carbonate, calcium carbonate, magnesium carbonate, aluminium hydroxide and magnesium hydroxide.

Antacids are available in different forms— effervescent tablets, chewable tablets and liquids.

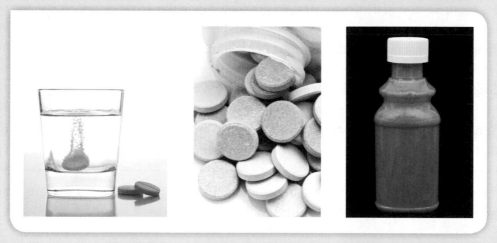

Your task as a group is to prove which of the provided samples of antacid are the most effective at neutralizing acid. As a group you must design and apply a procedure to test which antacid is the most effective. Remember that the description of your procedure must be complete, safe and detailed enough so that another person could use your procedure and reproduce your results.

You may want to use the entire amount of the sample, or your group may decide to use only part of it.

[Safety] Wear eye protection at all times. Dilute hydrochloric acid is an irritant: avoid contact with eyes, mouth and skin.

STEP 1 Discuss the potential ways the effectiveness of an indigestion tablet in neutralizing acid could be measured.

STEP 2 Write down the procedure you agree and have it approved by your teacher. Be sure you state a dependent variable and an independent variable.

STEP 3
a) Construct a data table to record your data.
b) Carry out your test.
c) Display the results in a suitable chart or graph.

STEP 4 **Conclusion and evaluation**

a) **Identify** the most effective antacid, and justify your choice.
b) Comment on how the results compare with the results of the other groups.
c) **Explain** how this procedure could be modified so that the most effective antacid could be used to test which acid from several unknown samples is the most concentrated.
d) **Evaluate** the procedure and state what improvements you would make if you were to carry it out again.

🌐 **GLOBAL CONTEXTS**
Scientific and technical innovation

🧠 **ATL SKILLS**
Information literacy
Collect and analyse data to identify solutions and make informed decisions.

🔗 **CHAPTER LINKS**
The causes of acid precipitation, and the reasons why even unpolluted precipitation has a pH of around 5.6, are discussed in Chapter 10 on consequences.

The problem of acid precipitation

Precipitation is considered acidic if the pH is less than 5.6. Acid precipitation is a serious problem in many parts of the world. Acid precipitation can cause long-lasting effects on ecosystems. Most freshwater fish species thrive in environments with a pH range from 6.5 to 8.2. They die if they are exposed to water with a pH lower than 5.0. Some lake ecosystems have a natural buffering system, which allows them to neutralize or counteract acid rain. This buffering action is caused by the presence of limestone, which consists of calcium carbonate. If an ecosystem does not have a natural buffering system such as limestone, the acid precipitation will not be neutralized and will harm the ecosystem.

 **Activity 7**   **Buffering capacity**

Imagine you have recently been employed by a local freshwater fish farm to monitor and analyse the effect of acid precipitation on the fish stock. Design a study to determine whether or not acid precipitation will affect the fish farm stocks.

Some questions to consider:

- What types of fish are raised in the fish farm and what are their pH tolerance levels?
- Does the local environment have a natural buffering capacity?
- What is the average amount of annual precipitation?
- What is the average pH of the precipitation in the area?
- What sources should you consider to collect your information?
- How much data would you need to collect?
- Who will you need to inform of your findings?

What other questions would be important to consider as you design your study?

GLOBAL CONTEXTS
Scientific and technical innovation

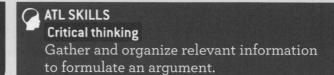
ATL SKILLS
Critical thinking
Gather and organize relevant information to formulate an argument.

Reflection on Topic 2

Indicators have an important function in the study of acids and bases: they tell us the specific pH of a solution. Some indicators change colour in the presence of an acid solution, and others in the presence of a basic solution. Some—like universal indicator and natural indicators like red cabbage juice—can function across the entire pH range. Indicators can also tell us when a neutralization reaction has occurred. Digital pH meters or probes give numerical readouts of the pH of solutions across the full pH range and are less subjective than interpretations of the colour change of an indicator.

- Comment on the precision and **sensitivity** of the use of indicators and pH probes to measure the pH of a solution.

- What function do indicators have in our everyday lives?

TOPIC 3

Function of the eye and spectacles

When light travels from one medium to another, it bends in a process called refraction. The relationship between the properties of the two media is an essential reason for this bending of light. In this topic, you will discover everyday applications for refraction, such as how the human eye functions, and the function of lenses in spectacles.

Light energy travels in straight lines. Light in a vacuum travels at a speed of 3×10^8 m/s. However, as the medium changes, so does the speed of the wave. In water, light travels at 2.56×10^8 m/s. In glass, light travels at 1.97×10^8 m/s. The speed of light is at maximum in a vacuum, and is slower in every other medium. Refraction occurs as light moves from one medium to another. The ratio between the speed of light in a medium versus its speed in a vacuum is called the index of refraction.

index of refraction $(n) = \dfrac{c}{v}$ where $c = 3 \times 10^8$ m/s,

v = speed of light in medium

The index of refraction of some common materials is shown in Table 13.1. Materials with a high index of refraction are referred to as optically dense.

Material	n	Material	n
vacuum	1	acrylic glass (polymethyl methacrylate)	1.49
air	1.0003	crown glass (used in lenses)	1.52
water	1.33	sodium chloride	1.54
ethyl alcohol (ethanol)	1.36	diamond	2.42
fused quartz or silica	1.46		
Pyrex®	1.47		

Table 13.1 Index of refraction of some common materials

When referring to refraction, it is always useful to specify the normal, which is an imaginary line perpendicular to the surface of the medium at the point where the incident ray hits the surface of the medium.

When light travels from a less optically dense medium to a more optically dense one (eg from air to water), the ray of light will bend towards the normal (Figure 13.3a) so that $\theta_1 > \theta_2$. When light travels from a more optically dense medium to a less optically dense one (eg from water to air), the ray of light will bend away from the normal (Figure 13.3b) so that $\theta_1 < \theta_2$.

Figure 13.3 Refraction of light rays occurs at the interface between air and glass

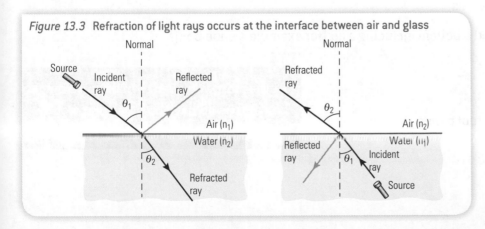

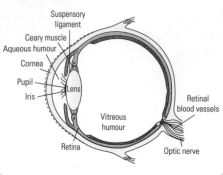

Figure 13.4 Cross-sectional diagram of the human eye

Suspensory ligament
Ceary muscle
Aqueous humour
Cornea
Pupil
Iris
Lens
Vitreous humour
Retina
Retinal blood vessels
Optic nerve

Function of the eye

The parts of the eye are shown in Figure 13.4 and their functions noted in Table 13.2.

Part	Function
cornea	Refracts light as it enters the eye.
iris	Controls the amount of light that is let into the back of the eye, by altering the size of the pupil (the central hole in the iris).
pupil	Changes size to adjust the amount of light let into the eye. In bright light, the size of the hole decreases to let in less light. In dim light, the size of the pupil increases to let in more light.
lens	Focuses light onto the retina.
retina	Senses light and creates nerve signals.
optic nerve	Transmits signals from the retina to the brain.

Table 13.2 The main parts of the eye and their function.

Our eyes function by focusing light entering the front of the eye (cornea and lens) onto the retina at the back of the eye (Figure 13.5). The retina is covered with billions of light-sensitive cells that communicate light patterns to the brain. The refraction of light through the cornea and lens is very important to achieve a sharp, focused image. To focus light reflected from objects at different distances, the shape of the lens can be changed by the ciliary muscles.

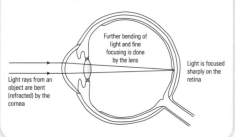

Figure 13.5 Refraction of light rays by the cornea and lens

Further bending of light and fine focusing is done by the lens

Light is focused sharply on the retina

Light rays from an object are bent (refracted) by the cornea

 Activity 8   Making a pinhole camera

STEP 1   Collect the following materials:

- box with four sides and a bottom including a lid (for example a shoe box)
- roll of aluminium foil
- sticky tape
- scissors
- tracing paper/greaseproof paper
- a convex lens.

TIP

At the start, you need to make sure the box and its lid have no holes that would let light into the box when the lid is on.

STEP 2 Cut two holes in opposite sides of the box so you are able to see right through. One hole should be circular and about 4 cm in diameter. The second hole should be rectang

STEP 3 Cut a square of foil making sure that it is bigger than the circular hole. Tape this to the inside of the box over the circular hole.

STEP 4 Make a very small hole in this piece of foil with the tip of a sharp pencil.

STEP 5 Cut a piece of the tracing paper/greaseproof paper making sure that it is bigger than the rectangular hole. Tape this on the inside of the box over the rectangle making sure it covers the entire hole. This will be your screen.

STEP 6 Tape the lid to the box. Your pinhole camera is now ready to use.

STEP 7 Point the foil end of your camera towards something bright, for example the window.

STEP 8 Observe the image on the screen. Record your observations—what did you see?

STEP 9 Observe and record what happens to the size, brightness and sharpness of the image when:

    a) the size of the small hole in the foil is increased
    b) you place more holes in the tinfoil (no more than five)
    c) you create a large hole in the foil of about 1 cm in diameter
    d) a convex (magnifying) lens is placed over this new large hole.

STEP 10 With the lens in place, observe and record what happens to the focus of the image when a small hole is placed in front of the lens.

Discussion questions

a) When you increase the number of holes, why do more images appear?
b) Why is the image upside down?
c) Why does the image become brighter but fuzzier when the hole gets bigger?
d) How does the lens cause the image to become more focused?
e) **Explain** how the properties of your pinhole camera can be linked to the functions of parts of your eye.

TIP

Drawing a diagram of your equipment set-up will help you explain why the image is upside down.

GLOBAL CONTEXTS
Scientific and technical innovation

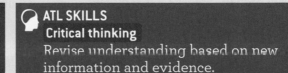
ATL SKILLS
Critical thinking
Revise understanding based on new information and evidence.

Correcting vision problems

Understanding refraction helps to correct vision problems caused when refracted light does not hit the retina in a focused beam, thus creating a blurry image. These problems are due to abnormalities in the anatomy of the eye—usually the shape of the eyeball (Figure 13.6). You may be born with an eyesight problem such as short sight (can see objects clearly when they are close, but have difficulty focusing on distant objects) or long sight (can see distant objects clearly, but have difficulty focusing on near objects). A type of long sight can also develop with age as the eye's lens becomes less flexible.

Figure 13.6 Cause of short sight and long sight

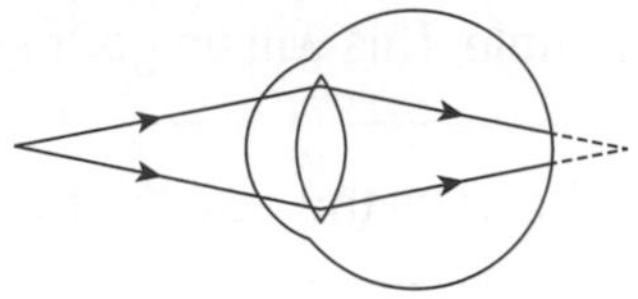

long sight – the eyeball is too short, so the lens focuses the sharpest image behind the retina

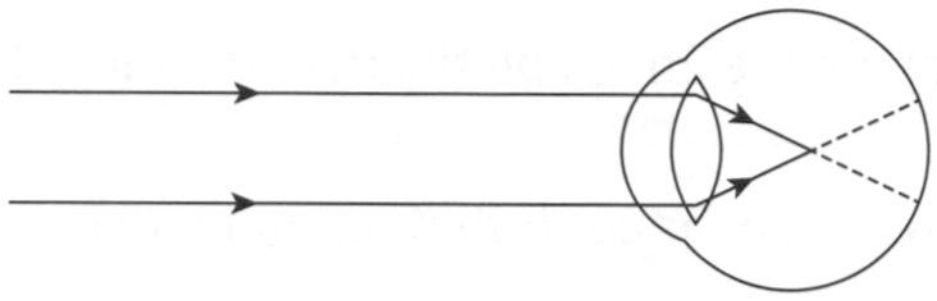

short sight – the eyeball is too long, so the lens focuses the sharpest image in front of the retina

QUICK THINK

Research how companies have been trying to make the function of lenses affordable to people in developing countries.

Long sight and short sight can be corrected by wearing glasses or contact lenses, which refract light prior to it passing through the cornea. This is to compensate for the error in our own eyes. Lenses for glasses are usually made of plastic or glass, because of the high index of refraction of these materials. The lenses are moulded to a specific shape in order to refract light at the correct angle into the patient's eye.

According to the World Health Organization, 1.3 billion people need glasses made to prescription to correct their sight. However, in many parts of the developing world, people do not have access to prescription glasses because they are expensive, and the professionals needed to prescribe them are not available.

 **WEB LINKS**

Try searching for "Joshua Silver" and "affordable glasses".

Activity 9   Refraction of light through different media

In this activity, you will measure the refractive index of different media.

Using the data in Table 13.1, predict which of the following materials will cause light to bend most when light enters the material:

- glass
- acrylic glass.

Give a scientific explanation for your hypothesis.

STEP 1 Collect the following materials:

- glass semicircular block
- acrylic semicircular block
- optional:
    - glass/acrylic rectangular block
    - optical blocks of acrylic/glass of different shapes and thicknesses
- ray box with power supply
- single slit
- protractor
- calculator.

[SAFETY] The ray box lamp will get hot enough to burn skin; take care. Do not use glass prisms with chipped or sharp edges.

STEP 2 Plug in the ray box. Be sure there is nothing else on the work bench as the ray box can get very hot. Use blinds/window covering to make the room darker than usual.

STEP 3 Adjust the ray box and slit so only one narrow beam of light is produced.

STEP 4 Place the semicircular block on a piece of paper. Draw a line perpendicular to the middle of the flat side of the block; this is the normal line. If semicircular blocks are not available, use a rectangular block instead.

STEP 5 Set up the ray box to shine a light ray through the block along the normal. Draw what you observe.

STEP 6 Move the ray box to shine light through the block at the following angles; 10, 20, 30, 40, 50, 60, 70. Draw what you observe in each case and record the angle of refraction in a table like the one below. Be sure to measure angles to the normal line and not the boundary line of the block. **Calculate** and record the values of sin i and sin r to two decimal places.

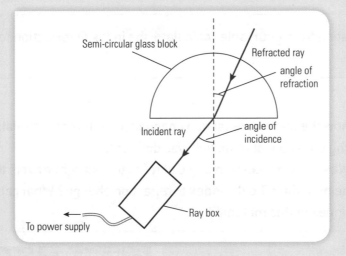

Angle of incidence (i)	Angle of refraction (r)	Sin i	Sin r	$n = \dfrac{\sin i}{\sin r}$
0°				
10°				
20°				
30°				
40°				
50°				
60°				
70°				
Mean value of n				

STEP 7 Repeat steps 5–6 using a glass semi-circular block instead of acrylic.

STEP 8 If possible, repeat steps 5–6 using glass or acrylic blocks of different shapes and thicknesses. There is a formula that links the angle of incidence, i, and angle of refraction, r, to the refractive index of each material:

$$\frac{n_2}{n_1} = \frac{\sin i}{\sin r}$$

where n_1 is the refractive index of the medium the ray is travelling from, and n_2 is the refractive index of the medium the ray is travelling into. Since the refractive index of air is very close to 1, the equation simplifies to

$$n = \frac{\sin i}{\sin r}$$

STEP 9 Using information from your table, **calculate** the index of refraction for acrylic and/or glass using your results.

Discussion

a) Describe how the angle of refraction changed for different materials (acrylic *vs* glass). **Explain** why the angle of refraction was different.

b) Optional: Was there a change in angle of refraction as light travels through different shapes of the same material? Did the index of refraction change? What can you conclude about the refractive index of this material?

GLOBAL CONTEXTS
Scientific and technical innovation

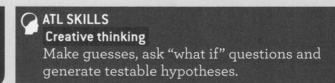
ATL SKILLS
Creative thinking
Make guesses, ask "what if" questions and generate testable hypotheses.

 **Activity 10**   Investigating a factor affecting the refraction of light

Now that you have completed one experiment on the refraction of light, you will plan your own experiment to explore the refraction of light in various media. Brainstorm the answers to these questions, and use this information to design your own experiment.

Brainstorming questions

a) What are some properties of white light?

b) What properties of light can you change in the experiment?

c) Would any of these properties affect refraction?

d) How does refraction work?

e) What apparatus can you change to affect refraction of light?

The following materials will be available for you to use:

- ray boxes
- coloured light filters
- power supply to make dimmer or bright light
- acrylic trays you can fill with various liquids
- glass or plastic clear shapes
- graph paper, protractors, metre rule.

[SAFETY] The ray box lamp will get hot enough to burn skin; take care. When working with liquids, do not handle power supplies or switches with wet hands.

Use your final idea to formulate a testable hypothesis and design a procedure for how the hypothesis could be tested. Remember to state the independent variable (the variable you will change), the dependent variable (what you will measure) and the control variables.

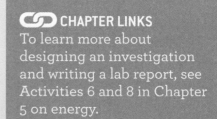 **CHAPTER LINKS**
To learn more about designing an investigation and writing a lab report, see Activities 6 and 8 in Chapter 5 on energy.

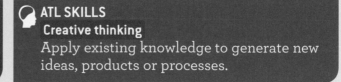

🌐 **GLOBAL CONTEXTS**
Scientific and technical innovation

💭 **ATL SKILLS**
Creative thinking
Apply existing knowledge to generate new ideas, products or processes.

Reflection on Topic 3

Refraction can be seen in many aspects of our daily lives. As you saw in Activity 9, refracted light can be manipulated by changing the medium in which the light travels.

- How was your pinhole camera different from the functions of your eye?

- How can the function of lenses be used to correct short-sightedness and long-sightedness?

- How does refraction explain the appearance of rainbows? From this, can you explain why cut diamonds sparkle with brilliant rainbow patterns, while uncut diamonds do not?

Summary

You have looked at examples of the function of individual components or objects that often perform together in a system. The examples you considered are sensory neurons and motor neurons in the nervous system, and the lens, cornea and retina in the eye. Can you think of other examples where components work together as a system to perform a particular function?

Other chapters have explored form and transformation. Scientific and technological advances enable researchers to develop innovative new devices by controlling and transforming the function of optical components. Agricultural or pharmaceutical biotechnology developments include transformation of organisms to perform different functions (eg genetically modified food, bacteria modified to produce human insulin).

It is important to understand the connection between form or structure and function, in order to be able to understand the behaviour of organisms, molecules or objects. Perhaps one day you will design new devices or products that take as their inspiration how components or systems in the natural world function.

Patterns

INQUIRY QUESTIONS	
	TOPIC 1 Classification
	■ **To what extent can fruit structures be used to infer relatedness in flowering plants?**
	■ **How can variation in giraffe coat patterns be explained?**
	TOPIC 2 The periodic table
	■ **What is a periodic pattern?**
	■ **What patterns are found in the periodic table?**
	TOPIC 3 Standing wave patterns
	■ **What patterns are made by reflecting waves?**
	■ **How can this knowledge of wave patterns be used to develop useful devices or create works of art?**

SKILLS

ATL

✓ Make inferences and draw conclusions.

✓ Organize and depict information logically.

✓ Build consensus.

✓ Process data and report results.

✓ Collect, record and verify data.

✓ Access information to be informed and inform others.

✓ Make connections between various sources of information.

✓ Revise understanding based on new information and evidence.

✓ Gather and organize relevant information to formulate an argument.

✓ Make connections between subject groups and disciplines.

Sciences

✓ Formulate a hypothesis using scientific reasoning.

✓ Organize and present data so that conclusions can be drawn.

✓ Make sketches of observations from an experiment.

✓ Create accurate, labelled scientific drawings.

✓ Interpret data gained from scientific investigations and explain the results using scientific reasoning.

✓ Plot scatter graphs and identify trends on graphs.

✓ Describe improvements to a method, to reduce sources of error.

✓ Make connections between scientific research and related economic factors.

OTHER RELATED CONCEPTS

Evidence   Form   Function   Models

GLOSSARY

Precision how close two or more repeated measurements are to one another.

Random error when measurement results vary in an unpredictable way. The effect of random errors can be reduced by making more measurements and calculating a new mean.

Systematic error when measurement results differ from the true value by a consistent amount each time a measurement is made.

COMMAND TERMS

Annotate add brief notes to a diagram or graph.

Apply use knowledge and understanding in response to a given situation or real circumstances. Use an idea, equation, principle, theory or law in relation to a given problem or issue.

Plot mark the position of points on a diagram.

Present offer for display, observation, examination or consideration.

Figure 14.1 Robert Millikan

Introducing patterns

In 1897, JJ Thomson discovered the first subatomic particle, a small particle with a negative charge we now call an electron. One of Thomson's students, John Townsend, developed a method for measuring the electron's charge but no scientist was able to get a very accurate result.

That was until a very patient and determined scientist named Robert Millikan tackled the problem (Figure 14.1). Millikan's method, a great improvement over previous attempts, is considered one of the most elegant experiments in the history of physics. His idea was to observe the motion of tiny electrically charged oil drops under gravity and in an electric field (Figure 14.2). Millikan realized that if you measured the total charge of a small but unknown number of electrons on the oil drop and repeated the experiment for a very large number of drops, you could find if these charges were all multiples of some smaller charge.

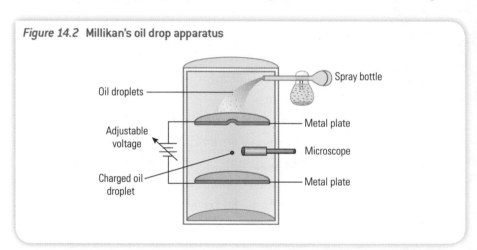

Figure 14.2 Millikan's oil drop apparatus

WEB LINKS

Search for "Millikan oil drop experiment" or browse the Chemistry videos on the network channel eLearnIn on YouTube.

The point here is not so much how Millikan obtained his data, but what he did with it. Millikan and his assistant Harvey Fletcher collected hundreds of data points over many years. His aim was to look through all the data for a pattern.

Millikan reasoned that if you collected enough data and knew the total charge on enough oil drops, there would be some number that would divide into all of these values of charge giving a whole number each time. This lowest common denominator would represent the charge on a single electron.

Millikan and Fletcher were rewarded for their relentless pursuit of a pattern by finding just such a number, 1.6×10^{-19} Coulombs.

In this chapter, you will explore the use of patterns in classifying organisms, in organizing elements in the periodic table and in predicting the behaviour of waves.

Classification

Pattern refers to the distribution of variables in time or space: sequences of events or features. An important part of the process of observation in science is to detect patterns in order to develop hypotheses. As an example, patterns can be used to decide if two species are related.

Taxonomy is the branch of science that focuses on placing organisms into named groups on the basis of shared characteristics. The scientific consensus is that organisms should be classified by how closely related they are. Closely related organisms are placed into the same groups. In order to determine relatedness, scientists use patterns.

Dichotomous keys

A dichotomous key repeatedly divides a group of objects into two mutually exclusive classes. That is, there are only two options for each division (eg yes and no). The objects are not necessarily divided into groups of equal sizes. For example, all the students in a class could be divided by gender and then by the courses they study (Figure 14.3). Dichotomous keys are sometimes organized as mutually exclusive numbered pairs of statements.

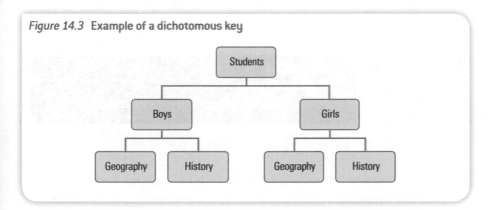

Figure 14.3 Example of a dichotomous key

TIP

A diagram such as Figure 14.3 can be created in Microsoft Word by inserting a SmartArt graphic. Alternatively, online applications such as bubbl.us can be used.

A dichotomous key is a way to organize and present information logically. It is possible that there is more than one way to logically structure the information.

A frequent misconception about the process and identification of species is that most new species are recognized in the field. In fact, many new species are identified using museum samples. There are vast collections of plants, insects and animals in museums and university laboratories around the world that represent a vast array of unstudied material.

Activity 1   Interpreting a dichotomous key

The following is a dichotomous key for a group of beetles.

1	wing cases with spots	ladybird
	wing cases without spots	2
2	elongated head	weevil
	head not elongated	3
3	long antennae	longhorn beetle
	short antennae	4
4	wing cases shorter than body	Oil beetle
	wing cases as long as body	Colorado beetle

Using the patterns described in the key, identify the following beetles.

A   B   C   D   E

⊕ **GLOBAL CONTEXTS**
Scientific and technical innovation

🗩 **ATL SKILLS**
Communication
Make inferences and draw conclusions.

Figure 14.4 Fruits like an apple develop from a single flower. There are never any petals left but the tube that extended from the ovary (the style, ending in the stigma) dries up and is left.

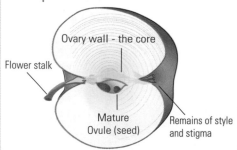

Ovary wall - the core

Flower stalk

Mature Ovule (seed)

Remains of style and stigma

From a botanical perspective, fruits are swollen ovaries that result after a flower has been pollinated. As the ovary develops, the other flower parts degenerate, but their remnants are visible on the mature fruit (Figure 14.4). In Figure 14.5 the sequence shows the relationship between development of flowers and fruit.

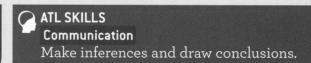
Figure 14.5 Pollination of a flower leads to fertilization, which leads to fruit formation

QUICK THINK

In everyday language the term fruit is often limited to sweet plant products that are seed-bearing. Following this pattern results in the grouping together of structures that are diverse in origin and excludes others that should be included. Research why strawberries are not viewed as "true fruits". Why are tomatoes, jalapeno peppers, cucumbers and squash not classified as fruits in everyday language?

 Activity 2 — Creating a dichotomous key of fruits

[SAFETY] Take care with sharp knives and always cut down on a board keeping fingers clear. Do not taste or eat any of the fruit.

STEP 1 Collect a range of fruits. Some recommendations are apple, lemon, pear, orange, tomato, pomegranate, plum, peach, grape and red pepper although other fruits can be used.

STEP 2 With each fruit, locate the position of the stalk where the fruit was attached to the plant and, at the opposite end, the scar that represents where the style was (Figure 14.4 above). Cut the fruit in half by following a line between the stalk and style remains.

STEP 3 Determine the number of sections in the ovary and where the seeds are located. You may have to make an additional parallel cut to locate the seeds.

STEP 4 Draw and **annotate** simple diagrams to illustrate the number of sections and the position of the seeds.

STEP 5 Examine more than one example of each fruit to determine if seed position and section numbers vary.

STEP 6 Based on the features you have observed, construct a dichotomous key. Deduce the taxonomical relationship between the fruits.

🌐 **GLOBAL CONTEXTS**
Scientific and technical innovation

🧠 **ATL SKILLS**
Communication
Organize and depict information logically.

 **Activity 3** — Analysing giraffe coat patterns

Pattern is obvious in the markings on different subspecies of giraffe (*Giraffa camelopardalis*). There are striking differences in the size, shape and arrangement of the dark patches. Up to nine subspecies of giraffe have been identified and they each have different patterns.

Pictured below is the coat pattern of four subspecies of giraffe.

A) West African

B) Rothschild

C) Reticulated

D) Masai.

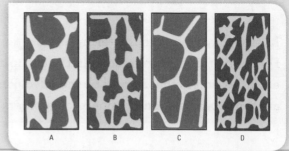

STEP 1 In your group, compare the coat patterns of the four subspecies of giraffe. Make reference to the regularity of the pattern.

STEP 2 Devise a method to determine the proportion of the coat area that is covered by patches.

STEP 3 Visit the database Arkive (www.arkive.org/) to research the geographic distribution of the different subspecies of giraffe.

 a) How do the habitats of the different subspecies vary?

 b) How does the vegetation vary?

 c) How does the temperature vary?

STEP 4 One explanation for the pattern of giraffe coat patches is that it serves as a form of camouflage. What testable hypothesis in relation to this explanation can you formulate?

TIP

When working in collaborative groups, it is important to begin with the habit of mind that "my own view does not exhaust the possibilities". It is possible that the pattern of the giraffe coat offers two advantages: heat regulation and camouflage.

STEP 5 An alternative hypothesis is that coat patterns are an adaptation for heat regulation. What predictions might you generate using this hypothesis? Discuss how you could test these predictions.

∞ LITERARY LINKS

The *Just So Stories for Little Children* are a collection of stories by the British author Rudyard Kipling. In the stories he offers explanations for such things as how the leopard got his spots, how the camel got his hump and how the elephant got his trunk. In everyday language, the phrase "just so stories" has come to mean unverifiable and untestable explanations for a phenomenon. It is tempting to devise any number of explanations for why giraffe spots patterns vary. However, scientists stick to hypotheses that can be verified.

But which is more fun to hear about during story time? You may be interested in exploring how explanations of natural phenomena vary from culture to culture.

GLOBAL CONTEXTS
Scientific and technical innovation

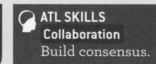
ATL SKILLS
Collaboration
Build consensus.

Patterns refer to the distribution of variables in time or space.

- Why is it important in science to observe and detect patterns?

- How can patterns be used to decide if two species are related?

TOPIC 2

The periodic table

The periodic table of elements is a chart that is arranged to show relationships among elements. Some of the elements have very similar characteristics while others do not. The periodic table is in the form that we know it today as a result of the work done by many different scientists over a number of years.

In 1829, Johan Dobereiner made one of the earliest attempts to group the known elements according to similar physical and chemical characteristics. He placed the elements into triads (groups of three), such as lithium, sodium and potassium. But as more elements were discovered they did not fit into this triad model.

In 1864, John Newlands was the next scientist to try to arrange the elements according to a pattern in properties. He arranged the known elements based on increasing atomic weight. He concluded that elements with similar properties occurred every eight elements—which he called the law of octaves. But, as with Dobereiner's model, the properties of newly discovered elements did not fit this arrangement.

In 1869, Dmitri Mendeleev also arranged the elements in order of their atomic weights. He arranged the 63 known elements into eight groups based on the patterns in their properties, such as density and melting point, so that elements with similar properties were in the same vertical column. What was different in his approach was that if necessary (because something did not fit), he simply left a space in his table and predicted that an element would be discovered to fill it. Mendeleev was able to predict the properties of the missing elements, several of which were soon discovered. Rather than trust only in the known elements, Mendeleev put his faith in the pattern of chemical properties resulting from scientific investigation and prediction.

The periodic table that scientists use today is based on Mendeleev's model.

WEB LINKS
If you are looking for a catchy new tune to download or for a new ring tone for your phone, you could try The Elements Song, which was first recorded in 1959. Search for "the elements song" on www.youtube.com.

A trend is a predictable change in a particular direction.

Using information from a database (see web links) you will investigate the relationships and trends in the following properties for the first 36 elements:

a) electronegativity
b) first ionization energy
c) atomic radius
d) ionic radius.

You will organize and **present** the data from the database so that conclusions can be drawn.

> **WEB LINKS**
>
> For the ionic radius visit www.lenntech. com and search for "periodic table". Visit periodictable.com for all other trends: once you are on this periodic table click on the element that you want to get information for and then click again on full technical data.

STEP 1 Define electronegativity, first ionization energy, atomic radius and ionic radius.

STEP 2 Use a spreadsheet to record and then **plot** a graph of the data found from the database. You need to decide how best to display the information. For example, will you plot all four sets of data on one graph or will you use separate graphs to display the four separate properties? Justify your choice of graph.

STEP 3 Describe and comment on the trends that you see and the element's position on the periodic table.

> **GLOBAL CONTEXTS**
> **Scientific and technical innovation**

> **ATL SKILLS**
> **Information literacy**
> Process data and report results.

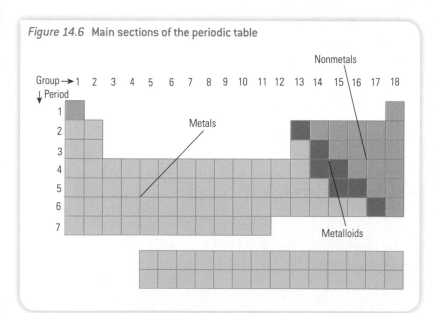

Figure 14.6 Main sections of the periodic table

As shown in Figure 14.6, the periodic table can be separated into two main sections—metals and non-metals—with metalloids bordering these two sections. Metals tend to have lustre (shine slightly), conduct electricity, be malleable and react with acids. Non-metals tend to be dull in appearance, be non-conductors of electricity and be brittle. Metalloids have properties of both metals and non-metals.

 **Activity 5**   Metal, non-metal or metalloid classification

You will investigate samples of unlabelled elements with respect to the following properties.

   a) Conductivity—through the use of a conductivity meter.
   b) Malleability—can you bend the element or is it brittle?
   c) Appearance—does it have lustre or not?
   d) Reactivity with acid—does a small sample of the element react when added to 1 cm^3 of 3M hydrochloric acid in a test tube?

[SAFETY] Wear eye protection and avoid skin contact with the acid.

STEP 1 Construct a data table to record your data.

STEP 2 Use the data that you collected to sort the elements into groups based on the similarities and differences in the physical and chemical properties that were observed.

🌐 **GLOBAL CONTEXTS**
Scientific and technical innovation

🧠 **ATL SKILLS**
Information literacy
Collect, record and verify data.

Groups and periods in the periodic table

Besides being divided into sections of metals, non-metals and metalloids, the periodic table is subdivided into a number of groups. The groups are the vertical columns and the elements within a group have similar characteristics (much the same way that family members have similar characteristics). They are often referred to as families.

There are 18 groups in the periodic table. The alkali metals are found in group 1, the alkaline earth metals in group 2, the halogens in group 17 and the noble gases in group 18. The transition metals are located between groups 3 and 12.

Periods are the horizontal rows in the periodic table and the atomic number increases as you go across a period. There are seven periods in the periodic table.

 **Activity 6**   Taking a closer look at the elements in a group

In this activity, you will use an online source of information to learn about the elements and the trends that exist in their families. Look at the videos for elements that are in the same group. Comment on and write down the patterns that you observe for elements that are located in the same group.

 **WEB LINKS**
You can watch videos of the elements on the periodic table at www.periodicvideos.com.

🌐 **GLOBAL CONTEXTS**
Scientific and technical innovation

💭 **ATL SKILLS**
Information literacy
Access information to be informed and
inform others.

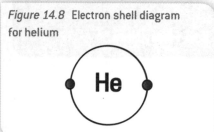

Figure 14.7 Electron shell diagram
for hydrogen

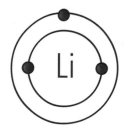

Figure 14.8 Electron shell diagram
for helium

Figure 14.9 Electron shell diagram for
lithium. The electrons are usually drawn at
the points on the compass, and only shown
in pairs once the shell has four electrons

Electron arrangements in atoms

One of the models for atomic structure discussed in Chapter 8 was
the Bohr model. In Bohr's model, electrons are found in shells at fixed
distances from the nucleus, each shell corresponding to a fixed energy
level. The seven periods or rows in the periodic table directly relate to
the seven possible shells or energy levels in Bohr's model of the atom.

In modern atomic theory, each shell can only hold a fixed number
of electrons. For the first 20 elements, the first shell can contain up
to two electrons and the second and third shells can each hold up to
eight electrons.

The electron arrangement of an atom specifies the number of electrons
in each energy level. An atom of hydrogen has one electron, found in
the first energy level. This can be displayed in the form H 1 or using the
electron shell diagram in Figure 14.7.

An atom of helium has two electrons, both found in the first energy
level. This can be displayed in the form He 2 or using the electron shell
diagram in Figure 14.8.

An atom of lithium has three electrons, which means it has
two electrons in its first energy level and one electron in the second
energy level. This can be displayed in the form Li 2, 1 or using the
electron shell diagram in Figure 14.9.

The electrons in the outermost shell (the valence electrons) are
responsible for chemical reactions and the bonding of atoms together.

🔗 **CHAPTER LINKS**
Chapter 8 on models describes how the electron shell model is
used to explain ionic and covalent bonding.

QUICK THINK

Search online for alternative
periodic tables. Choose one,
then research it and advocate
this model to the class.

Activity 7   Electron arrangement in atoms

Based on the pattern in electron arrangement that you have just seen in H, He and Li and the information given above, draw and write down the electron arrangement for the elements up to calcium, atomic number 20, using both formats.

Compare a graph of the first ionization energies for the first 20 elements, given opposite, with the electron arrangements you have drawn. Interpret the pattern shown in the graph and **apply** the electron shell model to explain this pattern.

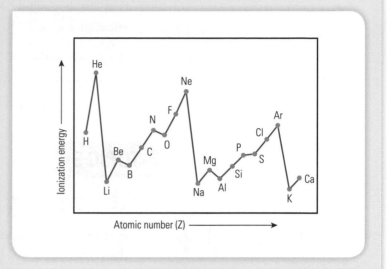

 GLOBAL CONTEXTS
Scientific and technical innovation

ATL SKILLS
Information literacy
Make connections between various sources of information.

Activity 8   The island of stability

In this activity, you will consider whether or not the hunt for "new" elements should continue, after looking at the process that scientists go through to create a new element at the end of the periodic table.

Scientists are continually trying to expand the number of known elements. Their process of discovery is based on fusing the nuclei of two elements to create a new element, which may last less than a millisecond before decaying radioactively.

WEB LINKS
You can follow the attempt by scientists such as Glenn Seaborg and Ken Moody to create element 114 in the laboratory by searching for "the island of stability element 114" on www.youtube.com.

a) Comment on the statement: "Transuranic elements (elements after uranium) should not be considered elements because they are not found naturally and can only be made in a laboratory."

b) Comment on the statement: "Research into the development of new elements is costly and should not be funded."

 GLOBAL CONTEXTS
Scientific and technical innovation

 ATL SKILLS
Critical thinking
Revise understanding based on new information and evidence.

The periodic table is central to the study of chemistry. The patterns that are displayed help to explain chemical reactions and properties of substances.

- Comment on the important role that patterns play in chemistry.
- Explain how the periodic table is a model for classifying and identifying the properties of elements.

TOPIC 3

Standing wave patterns

> **CHAPTER LINKS**
> Topic 3 in Chapter 8 on models has information on types of wave, interference and wave properties such as wavelength and amplitude.

Patterns help us understand and predict behaviour. In this topic, the pattern you will be studying is the standing wave pattern. Many everyday objects are designed to take advantage of the nature of standing waves. This pattern can be seen in vibrating strings on a guitar and any wind instrument. Standing waves in rooms can cause changes in sound levels depending on where the listener is standing. It is important information to have when designing auditorium acoustics.

Standing waves

Wave interference occurs when two waves meet while travelling along the same medium. When a wave is reflected back along the same path as the oncoming wave, the two waves that interfere are of the same frequency and equal amplitude. This type of interference can cause a pattern of alternating nodes and antinodes called a standing wave. The wave oscillates in space (Figure 14.10).

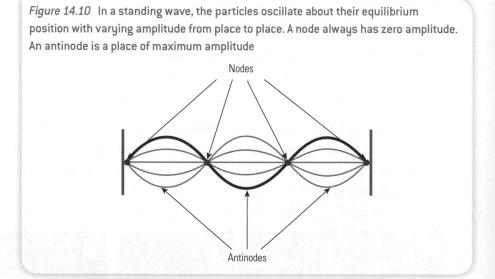

Figure 14.10 In a standing wave, the particles oscillate about their equilibrium position with varying amplitude from place to place. A node always has zero amplitude. An antinode is a place of maximum amplitude

Nodes

Antinodes

The nodes are a location where waves interfere destructively, causing no amplitude. In contrast, at the antinode there is maximum amplitude where incident and reflected waves interfere constructively.

In this activity, you will create a variety of standing waves with 2, 3 and 4 nodes.

STEP 1   Gather the following material:

- string or rope (greater than 2 m long).

STEP 2   Working with a partner on a smooth floor, stretch the string or rope between you.

STEP 3   Move your hand left and right in order to create standing waves following the patterns shown in the diagram. Only one student can move his or her hand, the other student must hold the rope or string stationary.

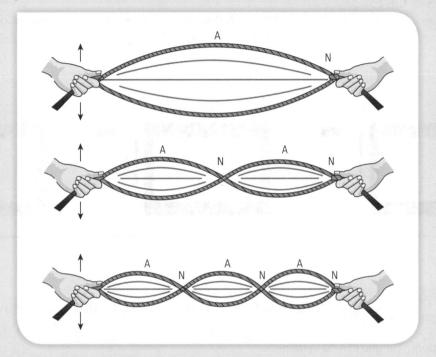

The different patterns of standing waves are called harmonics. The fundamental harmonic is where there is one antinode between two nodes, shown at the top of the diagram.

STEP 4   Sketch your observations.

STEP 5   Describe the amount of energy needed to create each standing wave with:

a) 2 nodes
b) 3 nodes
c) 4 nodes.

GLOBAL CONTEXTS
Scientific and technical innovation

ATL SKILLS
Information literacy
Collect, record and verify data.

Standing waves in microwave ovens

A microwave oven works by generating microwave radiation. Microwaves are a type of electromagnetic wave just like radio waves, infrared or visible light, but with wavelengths of 1–15 cm. They are produced by a device called a magnetron and enter the space inside the oven from a hole in the wall of the oven.

As shown in Figure 14.11, the microwaves are reflected from the other side of the metal oven back to the transmitter. The two waves are identical in frequency and interfere to create a standing wave pattern within the oven. There are antinodes where waves are at maximum, and nodes where the waves have no amplitude. When water molecules in the food absorb microwaves, they increase their kinetic energy. As the water molecules vibrate faster and faster, they collide more frequently with the molecules (water and other) around them. This transfers some of their energy to the surrounding food, raising its temperature.

QUICK THINK

How might the creation of antinodes affect the food being cooked?

Figure 14.11 Formation of standing waves in a microwave oven

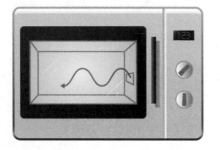  

 Activity 10   Visualizing standing waves in the microwave oven

This is an experiment you can do at home.

STEP 1 Collect the following materials:
- a large plate and a bowl
- four pieces of cold toast
- margarine
- a knife
- a ruler.

STEP 2 Remove the turntable from the microwave oven. This is very important. Place an upside-down microwave-safe bowl over the turning mechanism in the centre.

STEP 3 Arrange four pieces of cold toast in a square shape on a large plate so that the slices touch each other.

STEP 4 Cover the toast completely with a thick layer of margarine, including all the gaps between slices of bread.

STEP 5 Balance the plate on top of the upside-down bowl in the microwave.

STEP 6 Switch on the microwave oven at full power for 7–12 seconds until the margarine just begins to melt. More powerful ovens may need less time, so check after 5 seconds and again after another 5 seconds. Be very careful not to heat for too long.

STEP 7 You should see a series of parallel melted patches or lines separated by unmelted patches. Take the plate out of the oven.

STEP 8 Measure the distance in millimetres between two of these patches with a ruler. Measure from the centre of each patch to the centre of the next patch. Multiply by 2 and note down the value: this is the wavelength of the microwaves produced by your oven—it should be around 12–12.5 cm. Bring your data to class and compare with other students.

The distance between the melted spots represents half of the wavelength of the microwaves used. This is why you multiply the distance between the spots by 2 to get the wavelength. Look back at Figures 14.10 and 14.11 to check that the distance between two nodes or between two antinodes is half the wavelength of the wave that produces the pattern.

TIP

It is a good idea to repeat measurements to check if the values are grouped close together. If there is some variation in the measurements, the mean (or average) of the results is more accurate (see below) than the individual measurements.

Discussion

a) Why does a microwave oven use a turntable in order to cook food properly?

b) Although everyone used the same method, it is probable that there was a range of results in the class for the measured wavelength of a microwave. Suggest reasons for the differences. Are the measurements precise?

c) Suggest improvements that can be made to this experiment to increase the **precision** and accuracy of the results.

GLOBAL CONTEXTS
Scientific and technical innovation

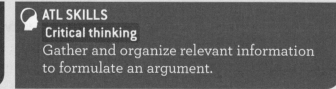
ATL SKILLS
Critical thinking
Gather and organize relevant information to formulate an argument.

In our everyday lives, the words accuracy and precision are used interchangeably. Yet, in science they have different meanings (Figure 14.12).

Accuracy is how close a measured value is to the known or actual value. If an object has a mass of 5.5 kg, and the scale reading is 5.2 kg, then the measurement is not very accurate.

Precision is how close two or more repeated measurements are to one another. If an object has a mass of 5.5 kg, and the scale readings are consistently reading 3.5 kg each time, then the measurements are very precise, but not accurate.

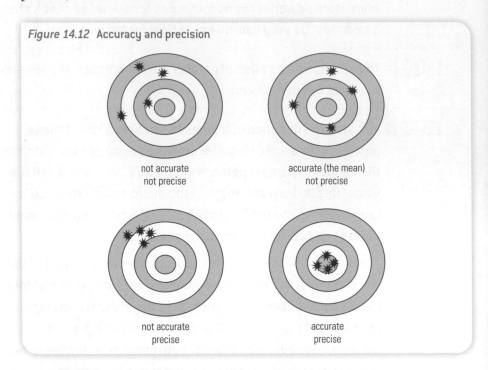

Figure 14.12 Accuracy and precision

not accurate
not precise

accurate (the mean)
not precise

not accurate
precise

accurate
precise

Understanding errors

Random errors in experimental measurements are caused by the inability of the experimenter in taking the same measurement in exactly the same way for every trial. This can be minimized by averaging over a large number of trials (such as a class set of data, completing many trials).

Systematic errors are due to problems that persist throughout the entire experiment, such as a zeroing error on an electronic scale. This error is apparent in all trials. This type of error may be difficult to detect if the experimenter is not careful to use the equipment properly.

Standing waves in sound

Loudspeakers create sound waves by turning an electrical signal into physical vibrations. The vibrating speaker moves air particles in front of the speaker back and forth, creating compressional or longitudinal waves. When these waves reach our eardrum, they eventually cause the movement of tiny hairs deep in the ear in an organ called the cochlea. The auditory nerve then relays the movement signals to the brain to be interpreted as sound. The "note" depends on the frequency at which the loudspeaker vibrates.

If two identical sound waves travelling in opposite directions interfere, they can interfere constructively and destructively to form standing waves.

By placing a fluid or light powder over the speaker, we can visualize the movement of the speaker. The fluid or powder is moved about as the speaker vibrates. The study of visible sound and vibrations is called cymatics.

Collect the following materials:

- large loudspeaker with the cone visible
- frequency generator
- clear plastic wrap
- beaker
- spoon
- 50 g corn starch (cornflour)
- 50 cm^3 water
- 100 g sodium chloride
- sturdy black poster board—cut to size to sit over the speaker with 2-inch overhang
- 12-inch dark-coloured plastic plate—should sit on top of speaker without being held
- thin metal sheet painted black—cut to size to sit over the speaker with 2-inch overhang.

WEB LINKS

There are good videos of this experiment on youtube.com. Search for "non-Newtonian cornstarch on speaker" and "Chladni patterns".

Part A

STEP 1 Make sure the signal generator is off. Turn the speaker over so the cone points upwards, forming a bowl. Loosely line the inside of the cone with clear plastic wrap. The plastic will hold the corn starch mixture, so make sure it does not have any holes and the corn starch mixture will not touch the speaker.

STEP 2 Connect the loudspeaker to the frequency generator.

STEP 3 In a bowl, mix 50 g corn starch with about 30 cm^3 of water, stir with a spoon. This mixture should be smooth and viscous. Add extra water if the mixture is not smooth, but do not make it too runny.

STEP 4 Pour the mixture into the speaker bowl.

STEP 5 Turn on the signal generator at a low frequency (a single, low-pitched note) and observe the mixture. Change the amplitude and frequency to see the changes in the vibrations.

STEP 6 Note the frequency and sketch a few of the patterns.

Part B

This time, the vibrating speaker passes its vibrations to a flat circular plate, which in turn moves very light solid particles.

STEP 1 Remove the plastic wrap. Place the thin metal sheet over the speaker. Be sure the sheet is horizontal. A tilted sheet will cause salt to spill.

STEP 2 Sprinkle 100 g salt to cover the sheet evenly in a thin layer.

STEP 3 Turn on the speaker at a low frequency, observe the salt.

STEP 4 Change the amplitude and frequency to see the change in standing wave pattern created by the sound waves. Sketch a few of the patterns. Add more salt if necessary as you increase the frequency.

STEP 5 Try using the plastic plate or a black poster board. Cut the poster board so that it can sit over the speaker bowl with at least 2 inches of extra cardboard extending around the speaker. Compare the patterns you see to the ones created with the metal sheet.

> **WEB LINKS**
> Consider creating art pieces using the patterns generated by the sound waves. See some examples at www.cymatics.co.uk.

 **GLOBAL CONTEXTS**
Scientific and technical innovation

 **ATL SKILLS**
Transfer
Make connections between subject groups and disciplines.

Reflection on Topic 3

You should understand the effect of reflecting waves in the creation of standing wave patterns. Understanding patterns allows us to predict the behaviour of this type of wave.

- How can you explain the patterns you observed in Activities 10 and 11?

- How can a guitar player change the frequency (pitch) of the sound produced by the guitar?

- Why does a piano not sound like a guitar even though they play the same notes?

Summary

As you have seen in this chapter, patterns have helped us to unravel some of the greatest scientific mysteries of the past century—including the fundamental questions of how species are related to other species, and how the 90 or so elements in the countless different molecules that make up our Universe can be organized into an elegant periodic table.

Scientists look for patterns or trends in results, and the trends can then be used to make predictions. In your science course, you will collect and process many sets of data, for sequences of events in time or the distribution of features over a distance, area or volume.

Hopefully, like the Nobel Prize-winning physicist Robert Millikan, you will find the patience and the wisdom to look through your data for the potentially ground-breaking patterns that may be waiting to be discovered.

Environment

INQUIRY QUESTIONS

TOPIC 1 Plant responses

- **How does mechanical stimulation affect plant growth?**
- **Are lichen patches larger on the tops of stone walls or on the sides of stone walls?**
- **What is the relationship between environmental variables and leaf width in a climbing plant?**
- **How does competition between seedlings affect their above-ground biomass?**

TOPIC 2 Gravitational force

- **How can we visualize the gravitational pull between bodies in the environment of space?**
- **What causes stronger gravitational forces around larger planets or stars?**

SKILLS

ATL

✓ Make guesses, ask "what if" questions and generate testable hypotheses.

✓ Interpret data.

✓ Draw reasonable conclusions and generalizations.

✓ Encourage others to contribute.

✓ Process data and report results.

Sciences

✓ Formulate a testable hypothesis and explain it using scientific reasoning.

✓ Design a method for testing a hypothesis, and select appropriate materials and equipment.

✓ Draw sketches of observations from an experiment.

✓ Explain how to manipulate the variables and how data will be collected.

✓ Process data and present it in an appropriate graph so that conclusions can be drawn.

✓ Interpret data gained from scientific investigations and explain the results using scientific reasoning.

✓ Discuss how well the data supports a conclusion.

OTHER RELATED CONCEPTS

Evidence   Interaction

GLOSSARY

Dependent variable the variable in which values are measured in the experiment.

Hypothesis a tentative explanation for an observation or phenomenon that requires experimental confirmation; can take the form of a question or a statement.

Independent variable the variable that is selected and manipulated by the investigator in an experiment.

COMMAND TERMS

Design produce a plan, simulation or model.

State give a specific name, value or other brief answer without explanation or calculation.

Introducing environment

Environmental factors include all the circumstances, objects, or conditions by which we are surrounded. The environment—in biology at least—is defined as all the biotic and abiotic factors that act on an organism, population or community and influence its survival, evolution and development. An environment can also refer to the physical conditions in your surroundings or in the universe as a whole. Physicists talk of the space environment, or a weightless environment.

There is a natural tendency to think big when we hear the word environment. You may have experienced discussions regarding the environment in terms of global issues. But environments can be very small as well, and one of the last great unexplored wildernesses may very well be the microbiome that exists within our own bodies.

Did you know that bacterial cells in or on your body outnumber your human cells by 10 to 1? It is surprising how little we know about our own cells. But being outnumbered is the very reason scientists have struggled with this. Filtering out the DNA that is human from that belonging to microbiota is no easy task and samples from faeces, skin, and saliva contain many thousands of different species.

In 2011, *Nature* published a study that takes the notion of the human microbiome a stage further. Researchers have been able to classify the kinds of bacterial community that live inside the human gut into three distinct types (called enterotypes).

The three main enterotypes are shown in Figure 15.1. Each enterotype is dominated by one bacterial genus. What has surprised scientists is that each type is associated with a specific ability to process certain nutrient types: *Bacteroides* to process carbohydrates, *Prevotella* to process proteins called mucins, and *Ruminococcus* to process mucins and sugars.

An even bigger surprise is that having one or the other of these bacterial types seems to be random. Around 40 people of different race, age, region, diet and gender were used in the study and there seems to be no connection of these factors to how exactly an enterotype develops.

Colonization of the gut by bacteria begins at birth, when a baby is exposed to bacteria in the external environment for the first time. Some scientists hypothesize that whichever bacterial community gets to this pristine and untouched environment first establishes itself and actively prevents the other two from taking over. Other scientists think that development of an enterotype may be linked to the immune system, with certain kinds of bacterial community able to coexist in certain conditions.

It is thought that intestinal bacteria can influence appetite and also the intestine's ability to extract energy from food. It is also possible that these bacteria stimulate our immune system and so play a role

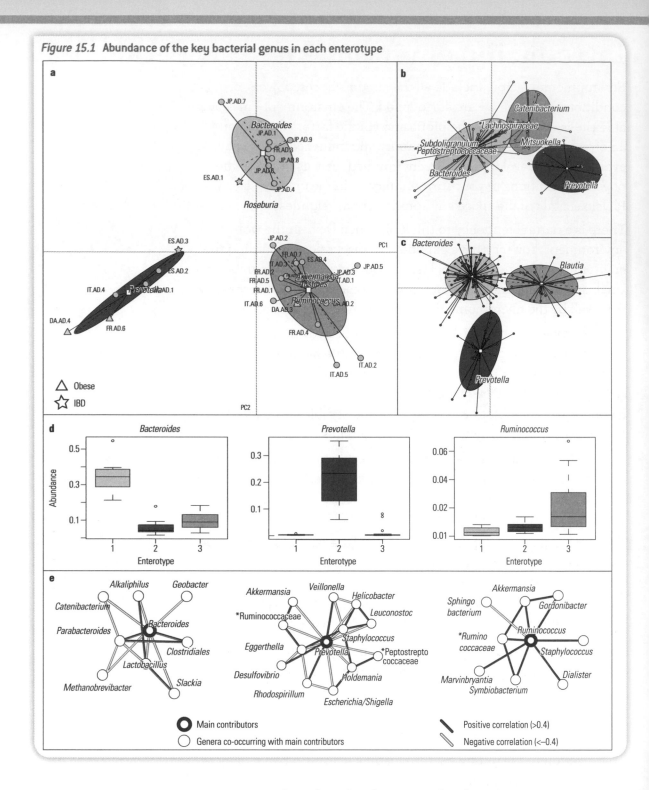

Figure 15.1 Abundance of the key bacterial genus in each enterotype

in autoimmune disorders (eg rheumatoid arthritis). In autoimmune diseases, the immune system attacks normal body tissue. Could knowing our enterotype affect the way we look at human health issues like obesity, type-2 diabetes, skin allergies, asthma and a range of metabolic and immunological disorders? Could this be a new tool to replace increasingly ineffective antibiotics in fighting human pathogens?

There are many other questions yet to be answered. Why these three types? What is special about them? Is it only these three, or are there more? How do we become colonized?

In this chapter, you will be considering how organisms respond to changes in the environment. You will also consider the larger environment of our Universe and how space itself responds to objects in its environment, warping its structure according to the laws of physics. Hopefully, your understanding of the concept of environment will become as broad as these topics have intended.

Plant responses

Plant growth and survival can be influenced by variables in the environment. These variables influence the responses, the development and the survival of individuals and the evolution of populations. If the variables are living, they are known as biotic factors. If they are non-living, they are known as abiotic factors. Organisms interact with the natural environment by transferring matter and energy.

 Activity 1   **Stem response to simulated wind**

Thigmotropism is the response of a plant to mechanical stimulation from an element in its environment. One of these responses has to do with growth patterns. Technicians who work with growing plants in greenhouses often note that the absence of wind leads the plants to develop weaker stems than plants exposed to wind. This can be explored using 2 cm radish seedlings. You could stimulate the plants by bending them repeatedly, each second for 30 seconds, using a pencil, at the same time each day.

STEP 1 Discuss with a partner possible ways that the treatment might be varied, and how stimulated and unstimulated plants may differ.

STEP 2 Using what you have discussed, suggest how your chosen treatment will affect the plant response. **State** a **dependent variable** and an **independent variable**, and say what the effect will be and why.

STEP 3 **Design** a procedure to test your **hypothesis**. What variables will you control? What measurements will you make, and how will you collect sufficient relevant data? A further question that can be explored is: do different species differ in their responses?

🌐 **GLOBAL CONTEXTS**
Scientific and technical innovation

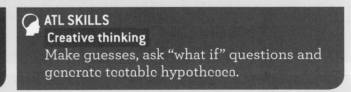 **ATL SKILLS**
Creative thinking
Make guesses, ask "what if" questions and generate testable hypotheses.

The photograph below shows a lichen patch growing on a stone. Lichens are plant-like in appearance but are close associations between two species: a fungus and a photosynthetic partner organism. The partner organism can be algae or it can be photosynthetic bacteria (previously known as blue-green algae).

How does the environment of a lichen influence its growth? Lichen patches are commonly found on stone walls. As the tops of stone walls receive more light, an experiment was performed to determine if there is any difference in the size of lichens growing on the top and side of a stone wall. Ten patches were measured in each location to determine the widest diameter of the patch. The data is presented below.

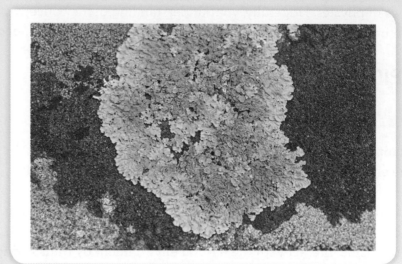

Surface	Diameter of lichen (mm)									
top	22	10	24	45	9	26	5	34	10	13
side	22	12	23	13	7	13	5	24	3	10

Source: Allott, A. 2014. *IB Biology Study Guide*. Oxford University Press.

Calculate the mean diameter for each of the two samples.

Discussion

a) Discuss whether the data supports the conclusion that there is a difference in patch size on the top and side of walls.

b) What are some possible limitations to this conclusion?

🌐 **GLOBAL CONTEXTS**
Scientific and technical innovation

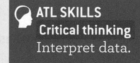
💭 **ATL SKILLS**
Critical thinking
Interpret data.

Most plant species produce more seeds than the environment can support. The result is a struggle for survival between individuals of the same species. The number of individuals that can be supported by an environment is known as the carrying capacity. The carrying capacity is a function of the environment and is determined by the light intensity, water and mineral availability.

 **Activity 3** — Competition

Competition for resources between individuals of the same species can be modelled in the laboratory by varying the number of seedlings sown in a pot.

STEP 1 Gather together the following materials:

- seeds
- soil
- medium-sized pots
- top pan balance.

[SAFETY] Always wash your hands after handling the soil and seeds.

STEP 2 Vary the number of seeds per pot from 4, 8, 16, 24 and 32. Space the seeds in a regular pattern. Once the seeds have begun to germinate, rotate the pots at regular intervals during the light period of the day. Grow the plants for one week after the appearance of true leaves.

After this length of time, cut the plants off at the point where they break through the soil. Measure the total wet biomass for the pot and the mean wet biomass of each plant. The wet biomass of plant is found simply by weighing the plant on a scale. Differences may be small so a scale that is sensitive to 0.01 g is recommended.

STEP 3 Construct a data table to record your data. When processing your data, select a method to indicate not only the mean value for each number of seeds, but also some indication of the variability of your data such as range bars, box and whisker plots or standard deviation, depending on what you have learned.

Analysis and conclusion

a) Does seedling density affect percentage germination?

b) Does seedling density affect above-ground total biomass? Does there appear to be a limit to the possible above-ground biomass per pot?

c) Does seedling density affect above-ground biomass per plant?

d) How do the plants grown in high-density conditions differ from those grown with less competition?

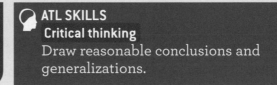

GLOBAL CONTEXTS
Scientific and technical innovation

ATL SKILLS
Critical thinking
Draw reasonable conclusions and generalizations.

Reflection on Topic 1

The environment affects the growth, development and responses of individuals. It also affects the evolution of populations.

- How does mechanical stimulation affect plant growth?

- How does the position of lichen patches affect their size?

- Which environmental factor affects the leaf size in a climbing vine?

- How does competition in the environment affect above-ground biomass?

Gravitational force

Ancient societies in Mesopotamia, Egypt and China all show evidence of having an interest in predicting future astronomical events such as eclipses and the motion of comets and the stars. We now know that the movements of the Moon, planets, comets and galaxies are due to the effects of gravity. This weak attractive force between objects is what holds planets in orbit and helps to create the structure of galaxies.

Physicists spend a lot of time trying to understand how and why objects move in the environment around us, whether on Earth or in our Solar System. Without an understanding of gravity, they would not be able to make technological advances in flight and space travel.

Physicists study closed systems in our environment. In a closed system, the only effects on the motion of objects are due to the forces of the objects interacting with each other. An example of this is the satellites that orbit the Earth, where the only external force acting on them is the gravitational attraction of the Earth.

In this topic, you will explore how a satellite stays in orbit, visualize the gravitational effect between the two masses as disturbing space–time around them and investigate the acceleration due to gravity.

Satellites

A satellite is any body that orbits a planet or star. The Moon is a satellite of the Earth. Earth also has many artificial satellites. Without the force of gravity, a satellite would continue to travel out into space instead of following an orbital path. To understand the force of gravity, we will look at the theories put forth by two great scientists, Isaac Newton and Albert Einstein. According to Newton's law of universal gravitation, any two objects with mass exert attractive forces on one another.

Imagine a cannon ball fired at high speed from a cannon at the top of a mountain. As the cannon ball travels through the air, it is pulled down by the force of gravity and falls to the ground. If the ball is fired at a higher speed, it falls farther away. The Earth is spherical not flat, so as the cannon ball falls to the Earth, the Earth is curving away. The faster the ball is fired, the further around the Earth it goes before falling to the ground. If the ball is fired at a fast enough speed, it travels in a circular path following the curvature of the Earth below it. We call this an orbit, and the cannon ball is now a satellite of the Earth (Figure 15.2).

CHAPTER LINKS
Chapter 4 on relationships has information about the mathematical relationships in Newton's law of gravitation.

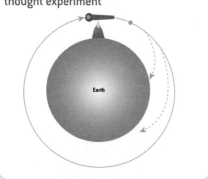

Figure 15.2 Newton's orbital cannon ball thought experiment

The Moon orbits the Earth. It doesn't fall into the Earth because of the high speed at which it travels around the Earth.

Returning to the cannon ball analogy. If the ball is shot out of the cannon at a much higher speed, it could potentially escape the gravitational pull of the Earth and continue out into space. The speed at which this occurs is called the escape velocity. Engineers calculating the launch speed of a satellite must keep this velocity in mind and not exceed it.

For more than 200 years, Newton's theory of gravity was the only accepted model to explain the motion of satellites. In the early 1900s, Einstein proposed a different explanation involving a change in what we call the fabric of space and time. According to Newton, time and space are individual concepts that do not change. Einstein suggested that both space and time are more fluid-like and can be altered by objects with large masses, such as stars and planets. It is easier to understand this if you think of space–time as the environment through which planets and stars exist. Einstein's theory of general relativity states that gravity is not a force between masses, as described by Newton, but is caused by a curve in the fabric of space–time in the presence of mass.

Massive bodies, such as stars, change their surrounding environment by creating a "dip" in space–time. Nearby moving objects are deflected from their straight-line paths to follow the local curvature of space–time (Figure 15.3). If such an object travels close enough to the star, the curvature of space–time forces it to travel around the star in a path we call an orbit. This is why planets orbit the Sun in our Solar System.

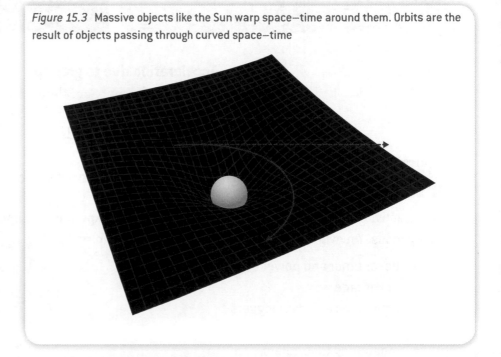

Figure 15.3 Massive objects like the Sun warp space–time around them. Orbits are the result of objects passing through curved space–time

Activity 4   **The relationship between mass and gravity in the universe**

You will use an analogy to visualize a gravitational field.

STEP 1   Collect the following materials:

- bed sheet or stretchy fabric (at least 1.5 m × 1.5 m)
- balls of various mass and diameter (eg tennis ball, basketball).

STEP 2   Group members each hold a section of the sheet so that the sheet is tight and flat and does not sag in the middle.

One person rolls the smallest, lightest ball across the sheet to someone else. Observe the path of the ball.

Note that this is how objects move in space when no forces act on them and space–time is not disturbed by any large mass.

STEP 3   Group members each hold a section of the sheet so that it does not sag in the middle.

Place the heaviest ball in the centre of the sheet.

One person rolls the smallest ball across the fabric to someone else.

Discuss what you notice about the path of this ball.

Sketch what you observe.

🌐 **GLOBAL CONTEXTS**
Scientific and technical innovation

🧠 **ATL SKILLS**
Collaboration
Encourage others to contribute.

Acceleration due to gravity

The force of gravity accelerates all objects down towards the Earth. For example, when you drop a ball or any other object, gravity accelerates it downwards.

 Activity 5   **Measuring the acceleration due to gravity**

In this activity, you will design your own inquiry-based experiment to investigate the acceleration of a falling mass. You will be provided with the following equipment:

- ticker timer and power supply
- ticker tape
- light gates and data loggers
- masses and mass hanger
- measuring tape/rulers.

🔗 **CHAPTER LINKS**
One method of measuring acceleration is described in Activity 9 in Chapter 12 on movement.

You may also use standard lab equipment such as a retort stand, string etc.

STEP 1 In your group, discuss the following questions:
- What variables are needed to describe the motion of a falling mass?
- What other adjustments to the experiment can be made?
- What variables must be controlled if measuring the acceleration due to gravity?

STEP 2 Formulate your own inquiry question and **design** a method for testing this question. Have your procedure approved by your teacher and make any changes that are needed.

STEP 3 Carry out your procedure.

GLOBAL CONTEXTS
Scientific and technical innovation

ATL SKILLS
Information literacy
Process data and report results.

The danger of space debris around the Earth

According to the website of the Union of Concerned Scientists Satellite Database, there are 1,235 operational satellites currently orbiting around the Earth (including launches up to August 2014). These include satellites used for military purposes, earth observation and communications. However, there are also many other human-made objects orbiting our planet, from chips of paint and broken pieces of booster rockets, to a screwdriver dropped by a spacewalking astronaut. In addition, satellite break-up due to collisions causes more space debris. The large number of small and large objects orbiting the Earth in low orbit environment is a real concern for current and future satellites.

The gravity assist technique

The curvature of space–time causes objects to orbit around a centre of mass. Understanding the behaviour of objects in the space–time environment gives us insight about how to successfully send spacecraft out to far-reaching parts of the Universe.

Reflection on Topic 2

Humans were fascinated by the effects of gravity long before it was explained by Newton's equation for the gravitational force between two masses. We know now that the movements of the planets and stars are all dependent on the effect of mass on the space–time environment.

There is more than one way to explain a phenomenon. Research and discuss this statement in relation to Einstein's and Newton's ideas about gravity.

QUICK THINK
Research and discuss the application and implications of human-built satellites in space.

WEB LINKS
Search for "space debris story" at www.spaceinvideos.esa.int/ to watch a video from the European Space Agency about the hazards to safe space navigation from space debris.

QUICK THINK
Research and discuss how understanding the behaviour of space–time has enabled scientists to gather information in outer space, looking specifically at the Juno spacecraft.

Summary

The living and non-living environments are systems, which can be described and explained by considering the interactions of the components of the system. The movement of planets and satellites depends on how the space–time environment is affected by the mass of objects such as stars. The survival and growth of organisms and populations depends on their interactions with the biotic and abiotic factors in the environment—that is, for example, the presence of other living organisms and the effect of physical conditions or climate in the habitat.

It is important to realize that there may be more than one way to explain a phenomenon and this is the exciting part of science, testing theories and looking for evidence in the environment around us.

WEB LINKS

Visit www. dystroy.org/ spacebullet/ for a gravity assist game—try your hand at launching a spaceship with the help of the gravitation field created by planets.

**KEY CONCEPT FOCUS
CHANGE**

INQUIRY QUESTIONS

TOPIC 1 Balance in biology

- How do disruptions to stable body temperature occur?
- What impact do environmental factors have on heart rate?
- What impact occurs when different trophic levels are removed from a model ecosystem?

TOPIC 2 Balance in chemical reactions

- How are mathematical relationships an integral part of chemistry?

SKILLS

ATL

✓ Gather and organize relevant information to formulate an argument.

✓ Interpret data.

✓ Make guesses, ask "what if" questions and generate testable hypotheses.

✓ Draw reasonable conclusions and generalizations.

✓ Understand and use mathematical notation.

✓ Process data and report results.

Sciences

✓ Formulate a testable hypothesis.

✓ Design a method and select appropriate materials and equipment.

✓ Explain how to manipulate variables, and how enough data will be collected.

✓ Organize and present data in tables ready for processing.

✓ Analyse data to draw justifiable conclusions.

✓ Draw conclusions, and explain these using scientific reasoning.

✓ Solve problems set in unfamiliar situations.

✓ Describe improvements to a method, to reduce sources of error.

OTHER RELATED CONCEPTS

Consequences Environment Evidence Models Transformation

GLOSSARY

Sensitive an instrument is very sensitive if it is able to respond to a small change in value.

Source of error a problem with an experimental method, the apparatus or ways in which the apparatus is used that cause measurements to be different from the true value.

COMMAND TERMS

Apply use knowledge and understanding in response to a given situation or real circumstances. Use an idea, equation, principle, theory or law in relation to a given problem or issue.

Estimate obtain an approximate value for an unknown quantity.

Outline give a brief account or summary.

Introducing balance

Balance is one of the few concepts in MYP Sciences that have specific definitions for specific disciplines.

In biology, balance has two meanings. First, balance is defined as the dynamic equilibrium that exists among members of a stable natural community (eg all the organisms living in a woodland habitat). A second biological meaning of balance is the regulation of the internal environment of an organism (eg its temperature or water content).

In chemistry, balance is defined as either a state of equilibrium (eg when the concentration of all reactants and products remains constant), or a stable distribution in a physical state (eg a stable but random distribution of sugar molecules throughout a glass of water if it is left undisturbed).

These ideas involve the idea of equilibrium. The equilibrium may be static or dynamic. In dynamic equilibrium, there is constant change of the individual members or components of the system, but the system is in balance and does not change overall.

Many natural systems exist in a state of balance or equilibrium and have the ability to return to a state of balance (which may not be the original state) following a disturbance.

It would be hard to think of a more profound disturbance than the disaster that happened at Chernobyl's nuclear power station in 1986. A chemical explosion caused a fire in one of the reactors. It burned for 10 days and blanketed thousands of square kilometres of the surrounding area with radioactive isotopes.

Particularly badly affected was a 10-square-kilometre area of pine trees directly surrounding the reactor. It was widely considered to be the most radioactively contaminated ecosystem in the world. Most of the trees were killed and the pine needles turned red-brown, resulting in it being called the Red Forest. The area was bulldozed and covered in sand to bury the dead trees (Figure 16.1).

But then something amazing happened in the 30 km exclusion zone around Chernobyl, where no humans are allowed to live or hunt. Plants began to grow. Insects returned, and then larger animals arrived—wolves, wild boar, elk, moose, deer, foxes, lynx, beavers, badgers, white-tailed eagles, nesting swans, cranes, black storks and great white egrets. All are thriving in a place where humans will not be allowed to live for at least another 20,000 years.

In this chapter, you will consider balance both biologically and chemically, and hopefully appreciate that—even if we cannot see it—stability and equilibrium are preferred conditions.

Figure 16.1 Forest in Chernobyl's exclusion zone

Balance in biology

Living organisms often possess mechanisms to maintain balance in their internal environment when external factors attempt to change that environment. Homeostasis is the regulation of the internal environment of an organism so that conditions remain relatively stable. Body temperature in mammals is an example of a variable that shows dynamic equilibrium by homeostatic mechanisms.

Figure 16.2 shows that the body's core temperature fluctuates around 37°C. What could cause the body temperature to increase or decrease above or below the range of homeostasis?

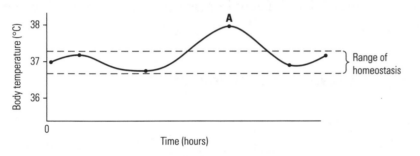

Figure 16.2 If body temperature becomes too high or too low, the body compensates so that temperature is returned to the normal range. After the disruption at A, the temperature returned to the range of normal homeostasis. This is called thermoregulation

INTERDISCIPLINARY LINKS

Language and literature

In his 1979 book *Gaia: A New Look at Life on Earth*, James Lovelock suggested that the Earth's biosphere (that is, all life on Earth) interacts with the environment to regulate temperature and chemical composition to keep the planet's overall conditions stable. The "Gaia hypothesis" has been influential but very controversial.

TIP

Balance does not mean that no changes occur. Homeostasis involves dynamic mechanisms that counteract changes to maintain the value of a variable at or near the set point.

Activity 1 — Exploring thermoregulation

Skin surface temperature can be monitored using skin surface temperature data loggers or by using liquid crystal strip thermometers, which are thermometers that are placed directly on the forehead.
A normal temperature is between 36°C and 37°C.

STEP 1 Measure the temperature of an extremity of your body (your fingers, toes or nose) using a skin temperature sensor or a strip thermometer. A skin sensor is preferred as the temperature of some extremities might be beyond the range of the strip thermometer. The skin surface temperature sensor is also more **sensitive** and responsive to slight temperature changes.

STEP 2 Put an ice cube in your mouth and push it up into the roof of the mouth, where it is closer to the hypothalamus, a structure within your brain that plays a role in temperature regulation. This is also an area that is well supplied with blood. Keep the ice cube there as long as possible, but if it starts to cause pain, move it away for a few seconds.

STEP 3 As soon as the ice cube has melted, take the temperature of the same extremity of your body as you did before putting the ice cube in your mouth.

STEP 4 **Outline** what your results suggest about thermoregulation.

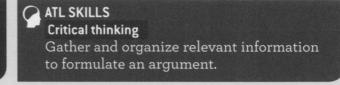

🌐 **GLOBAL CONTEXTS**
Scientific and technical innovation

🧠 **ATL SKILLS**
Critical thinking
Gather and organize relevant information to formulate an argument.

The diving reflex

When the faces of most mammals are submerged in water, their bodies respond with the involuntary "diving reflex". This involves slowing the heart rate to reduce oxygen consumption (bradycardia) and reducing the flow of blood to the skin surface.

In seals and whales, the reflex allows the animal to stay under water for some time despite an absence of oxygen intake. Diving triggers a new state of balance that is different from when the mammal can breathe air directly.

 Activity 2   **Interpreting data regarding the diving reflex**

The heart rate of an elephant seal and its corresponding dive depth were monitored for just over six hours. The data is shown in the graph.

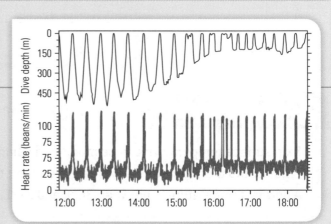

Questions

a) Determine the number of dives that occurred between 12:00 and 18:00.

b) Describe the pattern of dives that occurred over the day.

c) Find the depth of the deepest dive taken by the elephant seal. Measure as carefully as you can using a ruler.

d) **Estimate** the length of the longest dive taken by the elephant seal.

e) **Outline** the relationship between diving and heart rate.

 **GLOBAL CONTEXTS**
Scientific and technical innovation

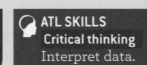 **ATL SKILLS**
Critical thinking
Interpret data.

 **Activity 3**   The diving reflex in humans

You are going to design an investigation related to the diving reflex in humans.

STEP 1   Discuss with a partner possible variables in humans that are affected by the process of diving, and which could be measured.

Using what you have discussed, formulate a hypothesis that could be tested. Include how diving affects this variable, and why.

STEP 2   Using what you have discussed, suggest how your chosen procedure will influence heart rate and why.

STEP 3   Design a procedure to test your hypothesis.

When designing your procedure, be sure to take the following into account:

a) Clearly state your dependent variable and independent variable.
b) List all of the other independent variables that should be held at a constant level (the control variables).
c) Describe how you will keep these other independent variables at a constant level.
d) List the equipment that you will need.
e) Describe instructions that another person could easily follow.
f) Be sure to include a design that involves enough subjects so that any patterns you observe have sufficient data to support them.
g) For each subject, your plan should involve taking repeat measurements.

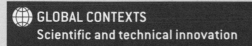 **GLOBAL CONTEXTS**
Scientific and technical innovation

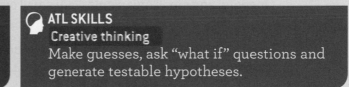 **ATL SKILLS**
Creative thinking
Make guesses, ask "what if" questions and generate testable hypotheses.

Balance in an ecosystem

Humans can disrupt ecosystems by selectively removing organisms from one trophic level. For example, wolves were hunted to extinction in many areas in Europe and North America leading to population explosions of the species that wolves preyed on. Reintroduction programmes have been undertaken in some areas but these are controversial.

Activity 4   Balance in a pond community

The dynamic equilibrium that exists among individual organisms in a stable natural community can also illustrate balance.

In this activity, you will explore how balance in communities is affected by initial conditions over time.

STEP 1 Gather together the following materials and apparatus:

    a) four 5 litre jars with lids (plastic film can be used if the jars do not have lids)
    b) a funnel
    c) gauze
    d) paper towels
    e) liquid plant fertilizer.

STEP 2 Obtain a sample of pond water and some of the material from the bottom of the pond. Stir the pond water and mud vigorously to mix it thoroughly.

STEP 3 Use a funnel to nearly fill one of the jars with the muddy water.

STEP 4 Nearly fill another of the jars, passing the water through a sieve to remove the largest organisms.

STEP 5 Nearly fill a third jar, passing the water through a piece of gauze to remove large and medium-sized organisms.

STEP 6 Nearly fill a fourth jar, passing the water through one or two sheets of paper towel to remove large, medium and small organisms. Very small organisms will be able to pass through the paper towel.

STEP 7 Loosely cover each of the model ecosystems that you have created with the lid.

STEP 8 Add a few drops of plant fertilizer to each of the model ecosystems.

STEP 9 Place all of the model ecosystems in a bright, warm place but not in direct sunlight. The model ecosystems may create some odour so if the weather is suitable, they can be kept outside.

STEP 10 Observe the growth of algae over time. Examine samples of the mud under the microscope occasionally to look for changes in terms of the variety and relative numbers of microscopic organisms.

STEP 11 Each filtering process might remove some combination of primary consumers, secondary consumers and tertiary consumers. This depends on the size of the mesh in the various materials. Make predictions of the effect on the model ecosystem of removing the following organisms:

    a) primary consumers   b) secondary consumers   c) tertiary consumers.

GLOBAL CONTEXTS
Globalization and sustainability

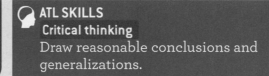
ATL SKILLS
Critical thinking
Draw reasonable conclusions and
generalizations.

Reflection on Topic 1

Systems involving living things often possess mechanisms to
maintain balance. This includes homeostasis in living things and the
balance that exists in stable ecosystems.

- What other examples can you think of where conditions
  inside the body must be maintained? Consider how the body
  responds to increase or decrease these levels when conditions
  are disrupted.

- A thermostat is used in swimming pools to keep the water
  temperature constant. The thermostat brings the water
  temperature back to a set value after it falls below this value.
  When the right temperature is reached, it switches the heating
  off again. Consider how well this analogy explains how the
  body maintains a constant temperature.

- Would the impact of removing all the organisms in one trophic
  level in an ecosystem be different for different trophic levels?

TOPIC 2

Balance in chemical reactions

Chemical reactions

When substances are mixed together, they may undergo a chemical
reaction. Evidence that a chemical reaction has taken place includes
production of a precipitate, production of a gas, a change in odour,
a colour change or a temperature change. If there is an increase in
temperature, the reaction is described as an exothermic reaction. If
there is a decrease in temperature, the reaction is described as an
endothermic reaction.

The products of a chemical reaction may have very different properties
from the reactants, but there is no change in mass when a chemical
reaction occurs. In 1789, Antoine Lavoisier discovered that mass is
neither created nor destroyed in chemical reactions. This is called the
law of conservation of mass.

Balancing chemical equations

A chemical equation is a shorthand way of describing what happens in a chemical reaction. These equations use symbols and formulas to indicate the substances involved in the chemical change.

Because of the law of conservation of mass, no atoms are created or destroyed—they are just rearranged. This means that in every chemical reaction, there must be an equal number of atoms of each element in the final products as there were at the start in the reactants.

A balanced chemical equation does not indicate the exact masses of substances involved in a chemical reaction. Instead, it indicates the relative amounts of each reactant and product.

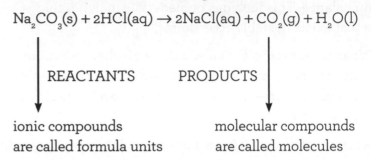

$$Na_2CO_3(s) + 2HCl(aq) \rightarrow 2NaCl(aq) + CO_2(g) + H_2O(l)$$

REACTANTS — ionic compounds are called formula units

PRODUCTS — molecular compounds are called molecules

This balanced chemical equation contains the chemical formulas for the reactants and products. It also indicates the relative amounts of each expressed as numerical coefficients in front of the formulas. If this equation were written in words it would say: "A formula unit of solid sodium carbonate reacts with 2 molecules of hydrochloric acid to produce 2 formula units of aqueous sodium chloride, 1 molecule of gaseous carbon dioxide and 1 molecule of liquid water."

The mole

Each chemical formula for a molecule or formula unit can be described as balanced in the sense that it has a neutral charge. For ionic compounds, the formula is written so that the positive ions and negative ions balance each other out.

Since atoms are extremely small, chemists needed a way to deal with the large numbers of atoms that are involved in any chemical reaction. This is provided by the SI system (*Le Système International d'Unités*). In this system, the base unit for an amount of substance is the mole, which has the symbol mol. A mole is a measurement term in the same way that a pair means two of something, and a dozen means 12 of something. A mole is a specific amount of something. That amount is 6.02×10^{23}.

1 mole is 6.02×10^{23} of anything
1 mole of donuts is 6.02×10^{23} donuts
1 mole of atoms is 6.02×10^{23} atoms
1 mole of molecules is 6.02×10^{23} molecules

WEB LINKS

To help you practise how to balance chemical equations take a look at these web links:

Beginner and brain boggling equations to balance at

http://funbasedlearning.com search for The Chembalancer.

An interactive balancing simulation at

https://phet.colorado.edu search for Balancing Chemical Equations.

A tutorial and practice with balancing chemical equations at

https://www.khanacademy.org search for Balancing Chemical Equations.

TIP

To help balance tricky equations, use these rules for the order in which to balance atoms:
- balance all atoms except hydrogen and oxygen
- balance hydrogen
- balance oxygen
- go back and recount the atoms to check the coefficients.

The number 6.02×10^{23} is called Avogadro's number. In 1811, Amedeo Avogadro contributed to the idea that a fixed amount of substance has a fixed number of atoms or molecules.

Much later in 1908, Jean Perrin was the first person to use this number for the number of particles in a mole.

Why is the mole significant? The mole is defined as the amount of a substance that contains as many elementary entities (atoms, molecules, ions, electrons, particles, etc) as there are atoms in 12 g of the isotope carbon-12. Another way to think about it is that the mass of 1 mol of atoms (6.02×10^{23}), expressed in grams, is equal to the relative atomic mass of the element. The mass of 1 mole of the substance is called its molar mass and has the units g/mol.

mass of 1 mole of oxygen (6.023×10^{23} atoms of oxygen) = 16.00 g
mass of 1 mole of gold (6.023×10^{23} atoms of gold) = 196.97 g
mass of 1 mole of water (6.023×10^{23} molecules of H_2O)

$$= (16.00 \text{ g}) + (2 \times 1.01 \text{ g})$$
$$= 18.02 \text{ g}$$
$$= 18.02 \text{ g/mol}$$

amount (in mol) = mass of substance (in g) / molar mass (in g/mol)

 Activity 5   **Dealing with very large amounts—the mole**

This activity will help you to understand and use mathematical notation.

STEP 1 Find food labels from various canned and packaged foods. Read the ingredients lists and copy down a total of 10 formulas or compound names. For each of the compounds you have listed, write down the correct formula and calculate its molar mass.

STEP 2 Your teacher will provide you with five bags that each contain a different substance. Each bag will be labelled with the name of the compound. Measure and record the mass of each chemical (assume the mass of the plastic bag is negligible) and calculate the number of moles and molecules that each bag contains.

STEP 3 You are asked to place 1.50 moles of zinc on a balance. How many grams of zinc will you need to place on the balance?

STEP 4 You are asked to place 6.02×10^{23} molecules of glucose on a balance. How many grams of glucose will you need to place on the balance?

STEP 5 Your teacher will provide you with a bag containing a sample of an element that is labelled with the number of moles that it contains. **Apply** your scientific knowledge to determine the identity of the element.

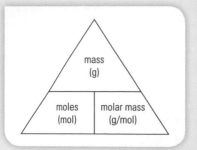

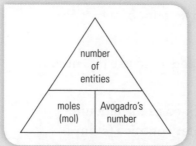

TIP

These equation triangles may help you to remember the relationship between the three variables, and to rearrange the equation.

GLOBAL CONTEXTS
Scientific and technical innovation

ATL SKILLS
Communication
Understand and use mathematical notation.

QUICK THINK

What data would you need to collect to calculate the number of water molecules that are in one mouthful of water?

 Activity 6   Determining the formula of a hydrate

When water is evaporated from an aqueous solution of a salt, water molecules often become incorporated into the crystals that form. These crystals appear to be dry, but when heated the trapped water molecules are released as water vapour. Salts with water molecules incorporated inside the crystal's structure are called hydrates. The trapped water is called water of hydration.

In this activity, you will be given a sample of an unknown hydrate along with a list of possible formulas for the hydrate in the form $M_aY_b \cdot xH_2O$ (s), where x is an integer. When the hydrate is heated, water is given off. The law of conservation of mass states that no atoms are created or destroyed in a chemical reaction. The balanced chemical equation for the reaction is:

$$M_aY_b \cdot xH_2O \text{ (s)} \rightarrow M_aY_b \text{(s)} + xH_2O\text{(g)}$$

Taken together, the equation and the law of conservation of mass indicate that the only atoms of hydrogen and oxygen are those in the hydrate and in the water given off. Hence, the mass of water that is lost when the hydrate is heated is the same as the mass of water that was originally in the hydrate.

Because you know the molar mass of the entire hydrate and the molar mass of just the water of hydration, you can calculate the percentage of water by mass in each of the possible formulas on the list provided by your teacher. You can then collect appropriate data to determine the mass of water lost when a known mass of the hydrate sample is heated. This information allows you to calculate the percentage of water by mass in the hydrate sample. You can compare this value to the values for percentage of water you calculated for each hydrate on the list. Thus, you can identify the formula of the unknown hydrate.

The apparatus available to you for this activity will be a test tube and test tube holder, a Bunsen burner, a ceramic tile and a balance. Your teacher may provide you with alternative apparatus.

[SAFETY] Wear eye protection. Heat the hydrate over a gentle heat and stop heating when the sample has lost all its water.

STEP 1 Calculate the percentage of water in each possible hydrate formula.

STEP 2 List what data you need to collect to determine the identity of your unknown hydrate, and construct a data table to record this data.

STEP 3 Heat your hydrate. How will you know when all the water of hydration has evaporated?

STEP 4 Using the collected data and the calculated percentage of water from each possible hydrate formula, identify the unknown substance.

STEP 5 Suggest how your results could have varied if some crystals had already lost their water of hydration.

STEP 6 Explain three possible **sources of error** in this experiment.

 **GLOBAL CONTEXTS**
Scientific and technical innovation

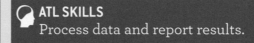 **ATL SKILLS**
Process data and report results.

Reflection on Topic 2

Balance is key to the study of chemistry. Balanced chemical equations allow chemists to estimate the amount of substances that are needed for a chemical reaction to take place. All sciences share mathematics, which can be considered the language of science and is a powerful tool to help with the development of laws and relationships. Some scientific explanations only exist in mathematical form. Laws like the law of conservation of mass are generally mathematical in form and can be used to calculate outcomes and to make predictions, such as how much product should be produced in a reaction.

- Explain why it is important to have a balanced chemical equation.

- Why is mathematics considered the language of science?

Summary

You have learned that balance is a fundamental condition in both chemistry and biology. You can use this knowledge to help you approach chemical reactions, and trust it to guide your mathematical approach to them. In biology, balance is necessary and, in many cases, an inevitable consequence of biological and natural systems.

Notes

Notes

Notes

Notes

Notes

Notes

Notes